中国土木工程学会
2021年学术年会论文集

中国土木工程学会　主编

中国建筑工业出版社

图书在版编目（CIP）数据

中国土木工程学会2021年学术年会论文集／中国土木工程学会主编. 一 北京 ：中国建筑工业出版社，2021.9
ISBN 978-7-112-26480-3

Ⅰ．①中… Ⅱ．①中… Ⅲ．①土木工程一学术会议一文集 Ⅳ．①TU—53

中国版本图书馆CIP数据核字（2021）第166494号

责任编辑：王砾瑶　范业庶
责任校对：姜小莲

中国土木工程学会2021年学术年会论文集
中国土木工程学会　主编

*

中国建筑工业出版社出版、发行（北京海淀三里河路9号）
各地新华书店、建筑书店经销
北京红光制版公司制版
北京建筑工业印刷厂印刷

*

开本：787毫米×1092毫米　1/16　印张：16½　字数：412千字
2021年9月第一版　2021年9月第一次印刷
定价：**145.00**元（含U盘）
ISBN 978-7-112-26480-3
（38047）

前　言

中国土木工程学会于 9 月 27 日～29 日在湖南长沙召开“中国土木工程学会 2021 年学术年会”。2021 年是中国共产党建党 100 周年，亦是“十四五”规划的开局之年，本次会议以“城市更新与土木工程高质量发展”为主题，邀请相关部门领导、院士、知名专家、科技人员和企业代表等参会，交流最新科研成果、探讨创新技术、展望未来发展趋势。旨在全面贯彻落实中央决策部署，紧密围绕国家发展战略，聚焦土木工程行业发展热点，推动我国土木工程向现代工业化、数字化、智能化转型升级，实现高质量发展。

本次会议的论文征集工作得到了广大科技人员的积极响应和踊跃投稿，共收到投稿论文 303 篇，经大会组委会组织专家审核，从中遴选出在理论上、技术上具有一定创新性和工程应用价值的论文 185 篇，汇编成《中国土木工程学会 2021 年学术年会论文集》（以下简称《论文集》）。另外，经专家审核、审稿，录用 5 篇发表在《土木工程学报》（增刊）。

《论文集》内容涵盖绿色低碳建设、城市更新、绿色建造、数字建造、智慧城市、现代桥隧、地下空间高效开发与利用、城市防灾减灾、土木工程高质量发展等方面。从低碳节能和便于查阅的角度出发，创新了出版形式，《论文集》纸质版只刊登论文目录及中英文摘要，论文全文以电子版形式出版。由于编纂时间较为仓促，难免有疏漏之处，敬请广大作者和读者予以谅解。

本次会议的组织召开及《论文集》的编辑出版工作得到了学会理事和常务理事、各分支机构、地方学会、会员单位以及本次会议承办单位中国建筑集团有限公司、中国建筑第五工程局有限公司的大力支持，在此一并表示感谢。

中国土木工程学会
2021 年学术年会组委会
2021 年 9 月于长沙

目 录

专题一 绿色低碳建设

专题二　城　市　更　新

专题三　绿　色　建　造

专题四　数　字　建　造

专题五 智 慧 城 市

专题六 现 代 桥 隧

专题七　地下空间高效开发与利用

专题八 城市防灾减灾

专题九 土木工程高质量发展

专题一　绿色低碳建设

盾构渣土复合双层免烧路面砖设计与工程指标测试

习智琴　李水生　阳　栋

（中国建筑第五工程局有限公司，湖南 长沙，410004）

摘　要： 以土压平衡盾构渣土为对象，结合粒化高炉矿渣，采用水玻璃与氢氧化钠溶液复合碱激发的形式，设计并研发了一种盾构渣土复合双层免烧路面砖。路面砖底层与面层高度比为 5∶1，通过将筛分出的 2mm 以下的盾构渣土与粒化高炉矿渣和液体激发剂混合作为底层料，保持了优良的蓄水功能；以碎石、砂、钢渣等作为面层，具有很好的透水功能。路面砖通过单向压力将底层和面层压制成型，其自然养护 28d 抗压强度达 51.8MPa，抗折强度为 9.1MPa，25 次冻融循环后的抗压强度为 46.4MPa，磨坑长度为 22mm，满足混凝土路面砖相关标准要求。底层物料采用 2mm 以下盾构渣土和粒化高炉矿渣，改善了渣土的粒径分布；将筛分出来的碎石、砂结合粒化高炉矿渣作为面层料，改善了路面砖的表面耐磨性能，添加钢渣进一步提高了耐磨性。路面砖底层与面层厚度比 5∶1 是可行的，成型压力较液固比对路面砖的抗压强度、抗折强度以及抗冻性能影响更大。

关键词： 地铁；盾构渣土；粒化高炉矿渣；钢渣；碱激发；免烧路面砖；工程指标

Double-layer unfired paving brick with shield muck: design principles and engineering index testing

Xi Zhiqin　Li Shuisheng　Yang Dong

(China Construction Fifth Engineering Division Co., Ltd.,

Changsha, Hunan 410004, China)

Abstract: In this paper, a novel double-layer unfired paving brick (SMPB) was designed and developed. The method consisted of combining earth pressure balance shield muck with ground granulated blast furnace slag (GGBS) and compound alkali excitation made from sodium silicate solution and sodium hydroxide solutions. The height ratio of the bottom to surface layer of SMPB was set to 5∶1. The bottom layer contained shield muck with size less than 2 mm, GGBS, and liquid alkaline activator with excellent water storage property. The surface layer contained crushed stone, sand, steel slag, GGBS, and liquid alkaline activator with good water permeability. The bottom layer and surface layer were pressed into brick shape through unidirectional pressure. After 28 days of natural curing, the unconfined compressive strength of the as-obtained SMPB was estimated to 51.8 MPa, flexural strength was 9.1 MPa, unconfined compressive strength after 25 freeze-thaw cy-

cles was 46.4 MPa, and length of grinding pit was 22 mm. These results met the requirements of relevant standards for concrete paving bricks. The bottom material composed of shield muck (size less than 2 mm) and GGBS improved the particle size distribution of the shield muck. The crushed stone and sand combined with GGBS enhanced the surface wear resistance of SMPB, and steel slag further improved the wear resistance. The height ratio of SMPB bottom to surface layer of 5 : 1 could feasibly be achieved. Also, the forming pressure of SMPB showed greater influence on its unconfined compressive strength, flexural strength, and frost resistance than liquid-solid ratio.
Keywords: subway; shield muck; GGBS; steel slag; alkali-activated; unfired paving brick; engineering index

施工升降机上屋面可周转门式附墙架施工技术

曹鸿皓　黄　俊　陈思远
（中国建筑第五工程局有限公司，湖南 长沙，410007）

摘　要： 株洲云龙万达综合体项目位于株洲市云龙区，商业广场区域高度23.95m，施工升降机为负责地上结构施工阶段材料的垂直运输工具。由于屋面层构架梁未完成施工，导致无法满足施工升降机移动至屋面层的条件。为了优化施工方法，项目部制作了一种新式的门式附墙架，并进行了基于力学分析与工艺可操作性的深化设计，以满足施工升降机安全运行到屋面的条件，达到节省人力与工期的目的。

关键词： 门式附墙架；有限元分析；施工技术

Construction technology of revolving door type attached wall frame on roof of construction lift

Cao Honghao　Huang Jun　Chen Siyuan
(China Construction Fifth Engineering Co., Ltd., Changsha, Hunan 410007, China)

Abstract: Zhuzhou Yunlong Wanda Complex Project is located in Yunlong District, Zhuzhou City. The height of the commercial square area is 23.95m. The construction lift is the vertical transportation tool responsible for the materials in the construction stage of the above-ground structure. Due to the unfinished construction of the frame beam of the roof floor, the conditions of moving the construction lift to the roof floor cannot be satisfied. In order to optimize the construction method, the project department made a new type of door attached wall frame, and carried out a deepening design based on mechanical analysis and process operability, in order to meet the conditions of safe operation of the construction lift to the roof, to achieve the purpose of saving manpower and time limit.

Keywords: door-mounted wall frame; finite element analysis; construction technology

高性能轻骨料预制大尺寸凸窗施工技术

吴　勇　周臻徽　赵天雪　孔德宇
（中建科技集团有限公司深圳分公司，广东 深圳，518000）

摘　要：采用800级圆球状高性能轻质陶砂，复合700级页岩陶粒，外掺改性聚丙烯纤维、聚羧酸减水剂，研究制备了一种新型高性能轻骨料混凝土，并用于大尺寸预制轻质凸窗的生产。对浇筑、收面工艺进行改进，减少了陶粒上浮比例及收缩裂缝。通过提前埋设斜撑连接件、放好边沿控制线、先装凸窗后绑墙柱钢筋、穿插现浇部分钢筋绑扎等施工措施，单个大尺寸预制凸窗安装时间在30min以内，效率可提高50%以上。
关键词：轻质陶砂；轻骨料混凝土；大尺寸；预制凸窗

Construction technology of high performance lightweight aggregate prefabricated large size window

Wu Yong　Zhou Zhenhui　Zhao Tianxue　Kong Deyu
(China Construction Technology Group Co., Ltd., Shenzhen Branch,
Shenzhen, Guangdong 518000, China)

Abstract: A new type of high performance lightweight aggregate concrete has been prepared by using 800 grade round spherical high performance lightweight ceramic sand, composite 700 grade shale ceramsite, mixed with modified polypropylene fiber and polycarboxylic acid water reducing agent, and used for the production of large-scale prefabricated light-weight bay windows. The process of pouring and finishing is improved to reduce the floating ratio and shrinkage crack of ceramsite. By laying inclined brace connection, putting edge control line in advance, first installing convex window and then binding wall column steel bar, inserting cast-in-place part steel bar binding and so on, the installation time of single large size prefabricated convex window is less than 30 min, and the efficiency can be increased by more than 50%.
Keywords: light ceramic sand; light-weight aggregate concrete; large size; prefabricated bay window

Mechanical properties of cemented sand and gravel materials based on artificial neural network

Fang Tao[1] *Wang Huilai*[1] *Li Fu*[1] *Li Zhaohui*[1] *Jin Guangri*[2]

(1. 3rd Construction Co., Ltd. of China Construction 5th Engineering Bureau, Changsha, Hunan 410007, China;

2. Department of Civil Engineering, Yanbian University, Yanji, Jilin 136200, China)

Abstract: The number of researches on the mechanical properties of cemented sand and gravel (CSG) materials and the application of the CSG Dam has been increased. In order to explain the technical scheme of strength prediction model about the artificial neural network, the study obtained the sample data by orthogonal test using the PVA fiber, different amount of cementing materials and age, and established the efficient evaluation and prediction system. Combined with the analysis about the importance of influence factors, the prediction accuracy was above 95%. This provides the scientific theory for the further application of CSG, and it will also be the foundation to apply the artificial neural network theory further in water conservancy project for the future.

Keywords: CSG (Cemented Sand and Gravel); artificial neural network; influence factors; strength; prediction model

吹砂回填珊瑚礁地基基础选型研究分析

陈　骏　庞海枫　周　翱　万　样　黄　飞
（中建三局第一建设工程有限责任公司，湖北 武汉，430040）

摘　要： 依托马尔代夫胡鲁马累7000套保障性住房项目，本文介绍了高层剪力墙结构在吹砂回填珊瑚礁地基上的基础选型过程。考虑珊瑚礁地质的特殊性，本文论证了CFG桩复合地基、钻孔灌注桩基础及钢管桩基础的技术可行性，并从成本、工期及组织难度等方面进行了比较研究。比选结果表明，CFG桩复合地基是该项目基础的最优选择。后续试桩结果也证明了CFG桩复合地基方案的施工可行性，项目基础选型过程及结果可为同类地基工程提供参考。
关键词： 吹砂回填；珊瑚礁；地基基础选型

Research and analysis on foundation selection of sand-dredger-fill coral reef

Chen Jun　Pang Haifeng　Zhou Ao　Wan Yang　Huang Fei
(China Construction Third Bureau First Engineering Co.,
Ltd., Wuhan, Hubei 430040, China)

Abstract: Based on the project of 7000 social housing units in hulhumale's phase 2, maldives, the paper introduces the foundation selection process of high-rise shearwall structures in sand-dredger-fill coral reef. Considering the particularity of coral reef geology, the paper demonstrates the feasibility of CFG pile composite foundation, bored pile foundation and precast tubular pile foundation, and makes a comparative study on cost, construction period and organization difficulty of three options. The result indicates that CFG pile foundation is the optimal choice for the project. The subsequent pile testing results have also proved the construction feasibility of CFG pile foundation. The process and results of foundation selection provide reference for similar projects.
Keywords: sand-dredger-fill; coral reef; foundation selection

新型绿色建筑材料——PVA 纤维增强土聚水泥的试验研究

卢　娟[1]　周红卫[2]　李　明[2]　周　游[2]

（1. 广州大学，广东 广州，510000；

2. 中建三局第三建设工程有限责任公司，湖北 武汉，430000）

摘　要：建筑业的迅速发展引发了一系列的环境问题，寻求绿色新型建筑材料已成为关乎民生与社会发展的重要任务。土聚水泥是一种三维立体网状结构的无机聚合物，生产过程无污染，但由于自身强度较低，导致其应用受到限制。本文将聚乙烯醇纤维（Polyvinyl alcohol fiber，简称 PVA 纤维）掺入土聚水泥，研究纤维掺量与纤维长度对其强度的影响。结果表明：随着 PVA 纤维掺量增加，土聚水泥抗压强度呈现先上升后下降的过程，在本文研究范围内，抗折强度还未出现下降趋势。不同长度的 PVA 纤维对土聚水泥强度的影响规律相近，但同等掺量的长纤维对土聚水泥的抗折强度提高作用更为明显。此外，PVA 纤维的掺入对土聚水泥早期强度影响不大。

关键词：土聚水泥；PVA 纤维；纤维掺量；纤维长度；强度试验

Experimental study on new green building material —— PVA fiber reinforced geopolymer

Lu Juan[1]　*Zhou Hongwei*[2]　*Li Ming*[2]　*Zhou You*[2]

（1. Guangzhou University，Guangzhou，Guangdong 510000，China；

2. China Construction third Engineering Bureau Group Co.，Ltd.，Wuhan，Hubei 430000，China）

Abstract：The rapid development of the construction industry has triggered a series of environmental problems. Seeking green new building materials has become an important task related to people's livelihood and social development. Geopolymer is an inorganic polymer with three-dimensional network structure. Its production process is pollution-free，but its application is limited due to its low strength. In this paper，PVA fiber was mixed with geopolymeric to study the influence of fiber content and fiber length on its compressive strength and bending strength. The results showed that with the increased of PVA fiber content，the compressive strength of geopolymeric increased first and then decreases. Within the scope of this study，the bending strength of geopolymeric had not shown a downward trend. The influence of PVA fibers with different lengths on the strength of

geopolymeric cement was similar, and the effect of long fibers with the same content on the flexural strength of geopolymeric cement was more obvious. The incorporation of PVA fiber had little effect on the early strength of geopolymeric cement.
Keywords: geopolymer; PVA fiber; content of fiber; length of fiber; strength test

基于模块化 MIC 整体卫生间的装配式集成设计分析

刘新伟　张亚东　王　豪　李　盼
（山东海龙建筑科技有限公司，山东 济宁，272000）

摘　要：以济宁任城区公厕改造项目为例，针对现阶段装配式建筑的结构建造方式与我国农村公厕存在的问题，提出了新型 MIC 集成房屋装配式建造技术。探究农村公厕建筑设计方案，反映农村特色并与周围环境相和谐，考虑社会、环境、节能环保等各方面因素，通过模块化 MIC 整体卫生间，营造出绿色节能、宜人使用的良好目标。运用装配式集成建造与结构分析技术，从设计源头出发，针对模块化 MIC 整体卫生间的连接节点构造形式进行创新，使 MIC 结构集成墙板、围护、装修、水电一次成型，现场快速连接，最终形成整体。探究各预制构件之间的组合形成新型的节点连接构造，综合分析其标准化、经济性与规模性。

关键词：装配式建筑；集成设计；公共卫生间；MIC

Analysis of assembly-type integrated design based on Modular Integrated Construction public toilet

Liu Xinwei　Zhang Yadong　Wang Hao　Li Pan
(Shandong Hailong Construction Technology Company Limited, Jining, Shandong 272000, China)

Abstract: Based on the public toilet renovation project in Rencheng District of Jining, aiming at the structural construction mode of prefabricated buildings and the existing problems of rural public toilets in China, a new MIC (Modular Integrated Construction) integrated building prefabricated construction technology was proposed. Exploring the architectural design scheme of rural public toilets, reflecting rural characteristics and harmony with the surrounding environment, considering social, environmental, energy conservation and environmental protection factors, and creating a good goal of green, energy saving and pleasant use through the modular MIC integral toilet. Using assembly integrated construction and structural analysis technology, starting from the design source, innovation is carried out for the modular MIC integral toilet connecting node structure form, so that the MIC structure integrates wallboard, enclosure, decoration, water and electricity forming once, and the site is quickly connected, finally forming the whole. And explore the combi-

nation of various prefabricated components to form a new node connection structure, comprehensive analysis of its standardization, economy and scale.
Keywords: prefabricated buildings; integration design; public toilet; Modular Integrated Construction

基于装配式全拼装公共卫生间的生产、施工技术分析

张宗军　王　健　张亚东　李　煦
（中建海龙科技有限公司，广东 深圳，518110）

摘　要：以济宁市任城区公厕改造项目为例，针对现阶段装配式建筑的工厂化生产、现场施工与建造方式进行分析，提出了全拼装墙板结构装配式绿色建造技术，探究农村公厕新型施工方案。通过ALC墙板结构进行既有公厕改造目标，完善农村公共卫生功能与空间关系，构建工厂内部流水作业，自动化、机械化程度高，水电管线、装修一体化，融入节能环保的理念。并通过合理编排的运输方案，搭配运输，将特殊构件拼装后整体运输，现场快速组装完成的新型绿色建造方式。工厂生产过程中按实际工程加以甄别材料类型、等级、尺寸、材料、钢筋信息、饰面、门窗等相关参数。通过绿色建造生产施工方式试图解决装配式墙体配套性、表面质量、开裂性、抗渗性等特性。同时探究全拼装结构的新型节点连接构造与绿色建造方式，极大程度节省人工成本和工期。

关键词：装配式建筑；生产施工；公共卫生间；绿色建造

Analysis of production and construction based on assembly fully assembled public toilet

Zhang Zongjun　Wang Jian　Zhang Yadong　Li Xu
(China State Construction Hailong Technology Company Limited,
Shenzhen, Guangdong 518110, China)

Abstract: Taking the renovation project of public toilets in Rencheng District of Jining as an example, this paper analyzes the factory production, site construction and construction methods of prefabricated buildings at the present stage, puts forward the new prefabricated construction technology of fully assembled wallboard structure, and explores the new construction scheme of rural public toilets. Through the ALC wallboard structure, the existing goal of public toilet renovation is to improve the rural public health function and spatial relationship, to build the internal flow operation of the factory, with a high degree of automation and mechanization, the integration of water and electricity pipelines and decoration, and to integrate the concept of energy conservation and environmental protection. And through the reasonable arrangement of transportation scheme, with transportation, the special components after the assembly of the whole transportation, the new construction method of site rapid assembly. In the factory production process according to the actual

project to screen material type, grade, size, material, reinforcement information, finishes, doors and windows and other relevant parameters. By means of production and construction, this paper tries to solve the characteristics of assembly wall, such as compatibility, good surface quality, not easy to crack and impermeability. At the same time, explore the new node connection structure of the fully assembled structure, which greatly saves labor cost and time limit.
Keywords: prefabricated buildings; production and construction; public toilet; green building

水泥固化重金属污染土的强度特性试验研究

林东海[1]　姜海峰[2]　白亚宾[1]　康晨雨[1]

（1. 中国建筑工程（香港）有限公司，中国 香港，999077；

2. 中国海外集团有限公司，中国 香港，999077）

摘　要： 我国土壤污染形势非常严峻，但相关法律法规出台较晚，土壤污染治理相关研究亟待开展。对处理重金属污染土，固化/稳定化技术具有极高的时间、经济等优势。相比于发达国家，我国相关研究仍处于初步探索阶段。本文主要以香港柴湾道391号泥土去污工程中重金属污染土为研究对象，着重研究水泥掺量及污染物浓度对固化污染土强度的影响。试验结果表明土体无侧限抗压强度随水泥掺量增加而线性增加，同时污染物浓度增加会少量增大土体无侧限抗压强度。

关键词： 固化/稳定化技术；无侧限抗压强度；水泥掺量；污染物浓度

Strength characteristics of heavy metal contaminated soils stabilized/solidified by cement

Lin Donghai[1]　*Jiang Haifeng*[2]　*Bai Yabin*[1]　*Kang Chenyu*[1]

（1. China State Construction Engineering（Hong Kong）Limited，Hong Kong 999077，China；2. China Overseas，Hong Kong 999077，China）

Abstract：The soil pollution situation in my country is very severe，but relevant laws and regulations were issued late，and research on soil pollution control needs to be carried out urgently. The solidification/stabilization technology has extremely high time and economic advantages for the treatment of heavy metal contaminated soil. Compared with developed countries，related research in my country is still in the preliminary exploration stage. This article mainly takes the heavy metal contaminated soil of the foundation project of Chai Wan Road in Hong Kong as the research object，and focuses on the influence of cement content and pollutant concentration on the strength of solidified contaminated soil. The experimental results show that the unconfined compressive strength of the soil increases linearly with the increase of cement content，and the increase of the pollutant concentration will increase the unconfined compressive strength of the soil by a small amount.

Keywords：solidification/stabilization technology；unconfined compressive strength；cement content；pollutant content

超高层高品质住宅人文健康与科技绿色集成产业化技术研究与应用——以深圳中海鹿丹名苑项目为例

刘 恋[1] 林 灏[2] 王 抒[1] 胡朔宾[1]
(1. 中国海外发展有限公司，广东 深圳，518000；
2. 深圳海智创科技有限公司，广东 深圳，519000)

摘 要： 在国民需求日益增长及国家发展绿色低碳战略目标背景下，产业化建筑是传统建筑产业转型升级的重要发展方向。中海通过深圳中海鹿丹名苑项目，研究与实践超高层产业化的同时，结合人文健康设计理念、利用创新科技与绿色节能技术进行集成升级，打造高品质住宅小区，改善并提高业主居住品质，具有较好的经济效益与社会效益，成为深圳市福利房住宅改造的典型案例，为后续同类型城市更新项目提供具有较高价值的借鉴和参考。

关键词： 城市更新；产业化；超高层；绿色节能；智能科技

Research and application of industrialization technology of Humanistic design concept and green building technology in super high rise high quality residential buildings ——take China Overseas Property-The Paragon as an example

Liu Lian[1] *Lin Hao*[2] *Wang Shu*[1] *Hu Shuobin*[1]
(1. China Overseas Land & Investment Ltd., Shenzhen, Guangdong 518000, China;
2. Shenzhen Haizhichuang Technology Co., Ltd., Shenzhen, Guangdong 519000, China)

Abstract: With the growing national demand and the development of green and low-carbon strategic objectives, Industrialized building has become an important development direction for the transformation and upgrading of traditional building industry. COB studied and practiced super high-rise industrialization through China Overseas Property-The Paragon, combining with the concept of humanistic health design, and using innovative technology and green energy-saving technology to upgrade. This project has become a typical case of the renovation of welfare housing in Shenzhen because of creating a high quality residential district, improving and enhancing the living quality of owners with good economic benefit and good social benefit. It can be used as a reference for the follow-up similar urban renewal projects with high value.

Keywords: urban renewal; industrialization; super high-rise; green energy-saving technology; Intelligent science and technology

重庆地区高星级绿色低碳住宅建设实践——重庆寰宇天下（B03-2/03）项目

黄荣波[1]　薛磊磊[2]　聂云茂[1]　文灵红[3]

（1. 中海地产重庆有限公司，重庆，400001；

2. 中国建筑科学研究院有限公司，北京，100013；

3. 中机中联工程有限公司，重庆，400001）

摘　要：不同建筑气候区，具有不同的建筑特点和节能要求，绿色建筑应因地制宜地提供健康、舒适、节能、低碳的建筑空间。重庆是典型的山地城市，夏季闷热、冬季湿冷，如何高度契合住户需求，为其提供高品质住宅空间的同时降低能源消耗，减少碳排放，是贯穿整个项目设计、建设和运营过程重点的问题。重庆寰宇天下（B03-2/03）项目结合重庆当地的自然气候条件、居民的生活习惯、当地节能要求以及中海地产多年的项目设计、建设及运营管理经验，从建筑方案、施工图、施工及运营管理的全过程进行管控。通过建筑物理模拟、第三方技术咨询、专家评审等多种途径，积极实践绿色建筑建设经验，项目获得国家三星级绿色建筑设计标识、重庆市居住建筑首个铂金级竣工标识和国家三星级绿色建筑运营标识。通过重庆寰宇天下（B03-2/03）项目的高星级设计、建设和运营的探索总结，其绿色建筑全过程管控的开发建设模式有较高的推广价值和参考意义。

关键词：绿色建筑；全过程管控；三星级；铂金级

Construction of Chongqing high-star residential building-practices in the project of Chongqing Huan Yu Tian Xia（B03-2/03）

Huang Rongbo[1]　*Xue Leilei*[2]　*Nie Yunmao*[1]　*Wen Linghong*[3]

（1. China Overseas Property Group Co.，Ltd.（Chongqing），Chongqing 400001，China；

2. China Academy of Building Research Co.，Ltd.，Beijing 100020，China；

3. CMCU Engineering Co.，Ltd.，Chongqing 400001，China）

Abstract：Different building climate zones have different building characteristics and energy-saving requirements. Green buildings should provide healthy，comfortable，energy-saving，and low-carbon building spaces based on local conditions. Chongqing is a typical mountainous city. The summer is sweltering and the winter is damp and cold. How to provide high-quality residential space while reducing energy consumption and carbon emissions is a key issue throughout the entire project design，construction and operation process. The Chongqing Huanyu World（B03-2/03）project combines Chongqing's local natural cli-

mate conditions，residents' living habits，local energy-saving requirements，and China O-verseas Land's years of project design，construction，and operation management experience. The whole process of operation and management is controlled. Through building physical simulation，third-party technical consultation，expert review and other channels，the green building construction experience is actively practiced. The project has won the national three-star green building design logo and the first platinum in Chongqing residential building Class completion mark and national three-star green building operation mark. Through the exploration and summary of the high-star design，construction and operation of the Chongqing Huanyu Tianxia（B03-2/03）project，the development and construction model of the whole process of green building management and control has high promotion value and reference significance.
Keywords：green building；whole process control；three Star；platinum

广州地铁东风站：方寸之间，回归绿色之本

唐 薇

（广州地铁设计研究院股份有限公司，广东 广州，510010）

摘 要：以“不以牺牲环境的代价换取经济上的发展，不以牺牲远期的利益来换取今天的发展”的可持续发展国家政策为前提，通过对广州地铁十四号线一期车站设计为切入点，以东风站为例进行较为全面而系统的介绍。在项目案例中以城市特有景观环境下，体会场地的历史人文或自然特性，设计具有地域特色的车站景观文化，力求建设可持续发展的、生态的、以人为本的以及可识别的地铁车站景观场所。以期对今后的地铁设计提供参考设计。

关键词：广州地铁；东风站；车站景观；文化场所

Guangzhou Metro Dongfeng Station：between the inches，return to the green

Tang Wei

(Guangzhou Metro Design & Research Institute Co. , Ltd. ,
Guangzhou , Guangdong 510010, China)

Abstract：On the premise of the national policy of sustainable development " not to sacrifice the environment for economic development, not to sacrifice the long-term interests for today's development", through the first phase of Guangzhou Metro Line 14 station design as a breakthrough point, taking Dongfeng Station as an example, a more comprehensive and systematic introduction is made. In the case of the project, under the unique landscape environment of the city, the historical and cultural or natural characteristics of the site will be experienced, and the station landscape culture with regional characteristics will be designed, striving to build a sustainable, ecological, people-oriented and identifiable subway station landscape place. In order to provide reference design for future subway design.

Keywords：Guangzhou Metro；Dongfeng Station；the landscape；culture place

绿建环评 BEAM Plus 评估体系在香港 DBO 项目中的应用研究

师 达 孔人凤 王欣欣

（中国建筑工程（香港）有限公司，中国 香港，999077）

摘 要： 目前绿色建筑已是全球发展趋势，全球已经有多个绿色建筑认证体系。中国香港地区的绿色建筑认证体系 BEAM Plus 经过多年的发展，已经具有成熟的技术体系及认证流程。本文将介绍香港 BEAM Plus 认证体系，并结合香港设计、施工及运营有机资源回收中心第二期项目，分析探讨为达到目标级别（铂金级）所采取的技术策略。通过分析总结中国香港地区市政项目的绿色建筑实践，为内地的绿色建筑发展提供参考和借鉴，为 2030 年二氧化碳排放达到峰值以及 2060 年前实现碳中和目标作出持续贡献。

关键词： BEAM Plus；绿色建筑；全生命周期；可持续发展

The study and application of BEAM Plus in Hong Kong DBO project

Shi Da Koong Lenfoong Wang Xinxin

(China State Construction Engineering (Hong Kong) Limited, Hong Kong 999077, China)

Abstract: Green building will become our future construction. Different rating tools for green buildings have been developed in various countries to assess the sustainability performance of a building. BEAM Plus is the Hong Kong's leading initiative to offer independent assessments of building sustainability performance. Since its first inception Beam Plus has developed a comprehensive set of performance criteria and a fair and objective certification process. In this paper we will introduce the assessments of BEAM Plus and the strategies used in the Design-Build-Operate (DBO) project of Organic Resource Recovery Center Phase 2 (OPARK2) to achieved the target of platinum status. The experience and outcome of applying green building strategies in municipal infrastructure projects will be analyzed and presented in this paper, in which the mainland can draw references and lessons for their development of green building as part of their effort to achieve peak carbon by 2030 and carbon neutral by 2060.

Keywords: BEAM Plus; green building; life cycle assessment; sustainable development

废弃矿山边坡绿化技术的研究

庞东喆　王　帅　郑艳超
（中建一局集团第三建筑有限公司，内蒙古 巴彦淖尔，015000）

摘　要：内蒙古地区是我国重要的经济增长极，矿山开采对区域建设提供了物资保障，但也留下了危岩、地面塌陷、基岩裸露、植被覆盖率下降、扬尘等严重的矿山地质和生态环境问题。为促进内蒙古地区协同发展，响应习近平主席"绿水青山就是金山银山"的号召，内蒙古地区废弃矿山的地质环境治理和生态修复形势急迫。以由中建一局负责实施的巴彦淖尔市乌拉特前旗矿山地质环境大坝沟西治理区为例，分析了废弃治理区存在的地质及生态环境问题，结合区域自然条件和生态环境，提出了矿山修复细分目标，并针对性地提出了成套技术方案，为内蒙古地区矿山修复提供了技术支撑，对区域矿山修复及环境改善提供了参考。

关键词：内蒙古地区；废弃矿山；地质环境治理；生态修复

Study on slope greening technology of abandoned mine

Pang Dongzhe　Wang Shuai　Zheng Yanchao
(The Third Construction Co., Ltd. of China Construction First Bureau Group, Bayannaoer City, Inner Mongolia 015001, China)

Abstract: Inner Mongolia is an important economic growth pole in China. Mining provides material support for regional construction, but it also leaves serious mine geological and ecological environment problems such as dangerous rock, ground collapse, bare bedrock, decreased vegetation coverage, dust and so on. In order to promote the coordinated development of Inner Mongolia and respond to President Xi's call that " green water and green mountains are golden mountains and silver mountains", the situation of geological environment treatment and ecological restoration of abandoned mines in Inner Mongolia is urgent. Taking dagouxi treatment area of mine geological environment in wulateqian banner of Bayannaoer City as an example, this paper analyzes the geological and ecological environment problems existing in the abandoned treatment area, puts forward the subdivision objectives of mine restoration combined with regional natural conditions and ecological environment, and puts forward a complete set of technical schemes, which provides technical support for mine restoration in Inner Mongolia, It provides a reference for regional mine restoration and environmental improvement.

Keywords: Inner Mongolia; abandoned mines; geological environment management; ecological restoration

河道整治技术研究
——以乌拉特前旗刁人沟河道整治为例

庞东喆　王　帅　王俊明　徐洪宇
（中建一局集团第三建筑有限公司，内蒙古 巴彦淖尔，015000）

摘　要：随着环境污染不断加重，以及人们环保意识的加强，河道综合整治引起人们高度重视。本文以中建一局三公司负责的刁人沟河道治理项目为例，对河道治理中地质环境和生态环境治理的具体措施与步骤进行研究，总结了刁人沟河道治理区有效的修复技术，达到了环境治理与生态修复的目的。通过项目实施可改善附近居民的生产和生活环境，消除开采区威胁周围居民及过往群众的安全隐患，保障人民生命财产安全，打造优美人居环境。为内蒙古等气候较干旱地区类似工程提供了借鉴。
关键词：河道现状；生态修复；河道治理

Study on river regulation technology

Pang Dongzhe　Wang Shuai　Wang Junming　Xu Hongyu
（The Third Construction Co. ，Ltd. of China Construction First Bureau Group，
Bayannaoer City，Inner Mongolia 015000，China）

Abstract：With the aggravation of environmental pollution and the strengthening of people's awareness of environmental protection，the comprehensive river regulation has attracted great attention. Taking diaorengou river regulation project in charge of the third company of China Construction First Bureau as an example，this paper studies the specific measures and steps of geological environment and ecological environment control in river regulation，summarizes the effective restoration technology of river regulation in diaorengou regulation area，and achieves the purpose of environmental control and ecological restoration. Through the implementation of the project，the production and living environment of the nearby residents can be improved，the potential safety hazards threatening the surrounding residents and the past people in the mining area can be eliminated，the safety of people's lives and property can be guaranteed，and a beautiful living environment can be created. It provides a reference for similar projects in arid areas such as Inner Mongolia.
Keywords：current situation of river channel；ecological restoration；river regulation

水肥一体化系统在乌拉山南北麓林业种植中的应用

任小龙　李天娇　王聚鹏　张　伟
（中建一局集团第三建筑有限公司，内蒙古 巴彦淖尔，015000）

摘　要：现代化林业对土壤水分、空气温度、湿度等作物生长的环境要求高，对灌水时间、灌水量、灌水部位、水肥营养供给等都有较高要求，滴灌智能化系统正是支撑现代化林业建设的一项基础性技术措施，亦是农业、林业走向现代化的必然趋势。水肥一体化系统是将灌溉节水技术、农作物栽培技术及节水灌溉工程的运行管理技术有机结合，同时集电子信息技术、远程测控网络技术、计算机控制技术及信息采集处理技术于一体，通过计算机通用化和模块化的设计程序，构筑供水流量、压力、土壤水分、作物生长信息、气象资料的自动监测控制系统，进行水、土环境因子的模拟优化，实现灌溉节水、作物生理、土壤湿度等技术控制指标的精确控制，从而将农业、林业节水的理论研究提高到现实的应用技术水平。

关键词：水肥一体化系统；人工造林；节水技术

Application of water and fertilizer integration system in forestry planting at the north and south foot of Wula mountain

Ren Xiaolong　Li Tianjiao　Wang Jupeng　Zhang Wei
（China Construction First Bureau Group No. 3 Construction Co.，Ltd.，
Bayannaoer City，Inner Mongolia 015000，China）

Abstract：Modern forestry has higher requirements for soil moisture，air temperature，humidity and other crop growth environment，and more requirements for irrigation time，irrigation amount，irrigation position，water and fertilizer supply. Drip irrigation intelligent system is a basic technical measure to support the construction of modern forestry，and is also the inevitable trend of Agriculture and forestry towards modernization. The water and fertilizer integrated system is an organic combination of irrigation water-saving technology，crop cultivation technology and operation management technology of water-saving irrigation project. It also integrates electronic information technology，remote measurement and control network technology，computer control technology and information collection and processing technology. Through the universal and modular design program of computer，it constructs the water supply flow，pressure，soil moisture，water quality control system The automatic monitoring and control system of crop growth information and meteorologi-

cal data can simulate and optimize the water and soil environmental factors, and realize the precise control of irrigation water-saving, crop physiology, soil moisture and other technical control indicators, so as to improve the theoretical research of agriculture and forestry water-saving to the practical application technology level.
Keywords: integrated system of water and fertilizer; artificial afforestation; water saving technology

绿色健康建筑体系下暖通专业解决方案

薛磊磊
（中国建筑科学研究院有限公司，北京，100013）

摘　要：在双碳目标下，作为我国高碳排放的建筑业，需要承担相应的减碳责任，因此建筑业需加快推动绿色建材、绿色建筑、近零能耗建筑发展进程。后疫情时代，人们对健康建筑的关注度也越来越高。本文主要介绍了暖通专业如何在保证舒适健康的生活环境下，降低建筑运行能耗，助力实现双碳目标。
关键词：绿色建筑；健康建筑；暖通；解决方案

Professional HVAC solutions under the green and healthy building system

Xue Leilei
(China Academy of Building Research Co., Ltd., Beijing 100013, China)

Abstract: Under the dual-carbon goal, as my country's high-carbon emission construction industry, it needs to bear corresponding carbon reduction responsibilities. Therefore, the construction industry needs to accelerate the development of green building materials, green buildings, and near-zero energy consumption buildings. In the post-epidemic era, people are paying more and more attention to healthy buildings. This article mainly introduces how the HVAC profession can reduce the energy consumption of building operation and help the dual-carbon goal while ensuring a comfortable and healthy living environment.
Keywords: green building; healthy building; HVAC; solution

碳达峰、碳中和背景下严寒地区超低能耗建筑规模化发展的价值分析——以中海地产呼和浩特市河山大观超低能耗项目为例

魏 纬 魏 刚 曲 斌 张瑞华 张 欢 金 阳 范 平
（呼和浩特市海巍地产有限公司，内蒙古 呼和浩特，010090）

摘 要：本文分析了建筑行业应对“30 60 碳目标”的碳减排路径，并提出超低能耗建筑是实现建筑业碳减排的基础和先决条件，分析了超低能耗减碳技术体系对建筑行业碳目标实现的价值，并以中海呼和浩特市河山大观为例，核算其碳减排量，论证了在严寒地区实现超低能耗建筑规模化发展对碳目标实现的巨大价值。

关键词：严寒地区；碳减排；规模化超低能耗建筑

Value analysis of large-scale development of ultra-low energy buildings in severe cold areas under the background of peak carbon emissions and carbon neutrality —— take the Heshan Daguan Project in Hohhot of China Overseas Property as an example

Wei Wei Wei Gang Qu Bin Zhang Ruihua Zhang Huan Jin Yang Fan Ping
（Hohhot Haiwei Real Estate Co.，Ltd.，Hohhot，Inner Mongolia 010090，China）

Abstract：This article analyzes the carbon emission reduction path of the construction industry in response to the "30 60 Carbon Target"，and analyzes the value of ultra-low energy buildings to achieve the carbon target of the construction industry，and puts forward that ultra-low energy consumption buildings are the basis and prerequisite for achieving carbon emission reduction in the construction industry，and analyzes the value of ultra-low energy consumption and carbon reduction technology systems to the realization of the carbon goals of the construction industry. Taking the Zhonghai Hohhot Grand View project as a sample，the carbon emissions are calculated and demonstrated that achieving large-scale development of ultra-low-energy buildings in severe cold regions is of great value to the realization of carbon goals.

Keywords：severe cold climate；carbon reduction；ultra low energy consumption building

专题二　城　市　更　新

城市老旧建筑混凝土屋盖改造组合加固施工技术研究

张　博　王　龑　温军伟　宿佩君
（中建三局第三建设工程有限责任公司，湖北 武汉，430000）

摘　要：建筑物随着时间推移因耐久性和使用功能发生改变，需对原有老旧建筑物进行结构改造加固以满足最新载荷和空间使用要求。在倡导绿色环保、节能减排的社会背景下，老旧建筑物的加固改造技术得到了广泛的推广及应用，多种加固技术的应用在日趋成熟的同时也在不断创新。本文通过对改造方案设计与研究，完成大跨度混凝土结构的托换与加固，通过现场施工过程数据的收集与成果总结等环节，提炼大跨度混凝土屋盖托换及组合预应力加固施工技术，为其他类似工程提供经验借鉴。

关键词：顶托卸载；组合加固；钢板张拉锚固系统；创新；改造

Combined reinforcement of old urban building concrete roof reconstruction research on construction technology

Zhang Bo　Wang Yan　Wen Junwei　Su Peijun
(China Construction Third Bureau Third Construction Engineering Co., Ltd., Wuhan, Hubei 430000, China)

Abstract: With the change of durability and function of buildings, as the time passing by, old buildings need to be reinforced to meet the latest load and space requirements. Under the social background of advocating green environmental protection, energy conservation and emission reduction, the reinforcement and reconstruction technology of old buildings has been widely promoted and applied, and the application of a variety of reinforcement technologies are becoming more and more mature and innovative. Completion the underpinning and reinforcement of the long-span concrete structure through the design and research of the transformation scheme, the construction technology of underpinning and combined prestressed reinforcement of large-span concrete roof is refined to provide experience for other similar projects through the data collection and achievement summary of the construction process.

Keywords: jacking unloading; combined reinforcement; tensioning anchorage system; innovation; reform

梁式框架受力结构钻石型蜂窝状的外凸型不规则幕墙施工技术应用

冯　颖　黄尚克　李书文　韦雨秀　何羽玲
（广西建工集团第五建筑工程有限责任公司，广西 柳州，545001）

摘　要：针对支承结构受力方式特殊，外凸型如钻石、蜂窝状的幕墙设计，为达到幕墙构造和施工要求，发明蜂窝状的外凸型不规则幕墙单体结构，在固有受力体系无法改变的情况下，运用几何原理将立体的幕墙外形分化为一个个平面形状，由点及面、由面及体的塑造立体，形成异形玻璃幕墙外形，最后详细介绍幕墙安装的施工过程。
关键词：幕墙；空间异性；蜂窝状；梁式受力；施工安装

Construction technology application of diamond-shaped honeycomb-shaped outer convex irregular curtain wall of beam frame forced structure

Feng Ying　Huang Shangke　Li Shuwen　Wei Yuxiu　He Yuling
(Guangxi Construction Engineering Group No. 5 Construction Engineering Co. , Ltd. , Liuzhou, Guangxi 545001, China)

Abstract: This article mainly elaborates on the special design of the supporting structure under the convex shape such as diamond and honeycomb curtain wall structure. In order to meet the design and construction requirements of the curtain wall structure, the honeycomb convex irregular curtain wall monomer structure was invented. Under the circumstance that the inherent force system cannot be changed, the three-dimensional curtain wall shape is divided into plane shapes by using geometric principles, and the three-dimensional shape is formed from points and faces, and from faces and bodies to form a special-shaped glass curtain wall shape. Finally, the curtain wall is introduced in detail. The construction process of installationis introduced.
Keywords: curtain wall; spatial heterogeneity; honeycomb shape; beam-like force; construction and installation

既有TOD地下室基于商业流线变化的结构加固探讨

刘慧明[1]　王　杨[2]
（1. 中海企业集团上海公司，上海，200092；2. 中海企业集团，广东 深圳，518000）

摘　要： 近年来随着绿色低碳建设的普及，越来越多的城市更新项目涌现。城市更新的商业项目会涉及商业流线的变化，而由此带来的结构问题也相对较多且复杂。本文基于一个实际工程项目，对商业改造过程中碰到的一系列结构问题，包括楼板的大开洞、新增柱网的转换、新老结构的搭接以及改造对于地铁的影响等问题，从理论计算及实际工程处理等方面进行了探讨，并进行了归并和总结，最后又对此类项目的施工方式提出了一定要求，由此得到的一些相关结论可供类似城市更新项目参考。
关键词： 城市更新；结构加固；结构转换；TOD

Discussion on structural reinforcement of existing TOD basement based on change of commercial streamline

Liu Huiming[1]　*Wang Yang*[2]
(1. China Shipping Enterprise Group Co., Ltd., Shanghai, 200092, China;
2. China Shipping Enterprise Group Co., Ltd., Shenzhen, Guangdong 518000, China)

Abstract: In recent years, with the popularization of green and low-carbon construction, more and more urban renewal projects emerge. The commercial projects of urban renewal, for example, will involve the change of commercial flow line, and the resulting structural problems are relatively more and more complex. Based on an actual engineering project, this paper discusses a series of structural problems encountered in the process of commercial transformation, including the large opening of floor, the conversion of new column network, the overlapping of new and old structures, and the influence of transformation on subway, and summarizes them from the aspects of theoretical calculation and practical engineering treatment, Finally, the paper puts forward some requirements for the construction methods of such projects, and some relevant conclusions can be used as reference for similar urban renewal projects.
Keywords: urban renewal; structural reinforcement; structural transformation; TOD

UHPC无损加固历史砖墙单面模板体系研究

徐　玉　严凯开
（上海建工二建集团有限公司，上海，200082）

摘　要：以黄浦区160街坊保护性综合改造项目为例，通过数值分析与现场试验对UHPC无损加固历史砖墙单面模板体系展开研究。提出型钢斜撑-模板体系与钢管扣件-模板体系，并建立有限元模型分析两种单面模板体系在UHPC浇筑侧压力作用下的受力变形；通过计算对单面模板体系的支脚强度与墙体的抗倾覆性进行验证；针对UHPC的材料特性与历史砖墙的特点，制定单面模板体系的施工工艺；通过现场试验，验证型钢斜撑-模板支撑体系作为UHPC加固历史砖墙单面模板体系的可行性，支撑体系无明显变形，加固面层光滑平整，施工精度与强度均满足设计要求，为UHPC无损加固历史砖墙单面支模体系在实际工程应用中提供参考依据。

关键词：UHPC；历史砖墙；无损加固；单面模板体系

Research on the single-sided template system to reinforce non-destructively historical walls by UHPC

Xu Yu　Yan Kaikai
（Shanghai Construction Engineering Second Construction Group Co.，Ltd.，
Shanghai 200082，China）

Abstract：With the background of the comprehensive renovation project of Huangpu 160 area，a single-sided template system to reinforce non-destructively historical walls by UH-PC was studied through numerical analysis and field test. Proposed the supporting system of section steel and steel pipe and established a numerical calculation model to analyze the force and deformation of the two single-sided template systems under lateral pressure；verified the strength of the legs of the single-sided template system and the overturning resistance of the wall through calculations；considering the material characteristics of UHPC and the characteristics of historical brick walls，formulated the construction technology of the single-sided template system；Through field tests，the beam support system is feasible，the support system was less deformed，the reinforcement surface was smooth and flat，and the accuracy and strength meet the design requirements，the conclusion provided a reference for the UHPC non-destructive reinforcement of the historical brick wall single-sided formwork system in practical engineering applications.

Keywords：UHPC；historic brick wall；non-destructive reinforcement；single-sided template system

既有历史保护建筑群上的大跨度异形曲面钢结构顶盖施工技术研究

郭雨棠
（上海建工二建集团有限公司，上海，200090）

摘　要：南京东路179号街坊成片保护改建工程项目在四幢既有历史保护建筑改造的基础上在其上空约22m处安装一座全覆盖大跨度异形曲面钢结构玻璃顶盖。通过对该工程的重难点分析，优化施工方案，控制关键施工工艺，合理安排施工流程，保证了该工程的施工质量，取得了良好的社会效益和经济效益，为同类型的项目施工提供了宝贵的施工经验。

关键词：既有历史保护建筑；大跨度；异形；曲面；钢结构施工

There are large span special-shaped curved surfaces on the historic conservation buildings research on construction technology of steel structure roof

Guo Yutang
（Shanghai Construction Engineering Second Construction Group Co.，Ltd.，Shanghai 200090，China）

Abstract：On the basis of the reconstruction of four existing historic buildings，a full-covering large-span special-shaped curved steel glass roof is installed at 22m above the block protection and reconstruction project of No. 179 East Nanjing Road. Through the analysis of the key and difficult points of the project，the construction scheme is optimized，the key construction technology is controlled，the construction process is arranged reasonably，the construction quality of the project is guaranteed，good social benefits and economic benefits are obtained，and valuable construction experience is provided for the construction of the same type of project.

Keywords：existing historic buildings；large span；abnormity；surface；construction of steel structure

专题三　绿　色　建　造

埃及 CBD 项目筏板超长对拉螺杆侧模系统的设计与施工

郝建兵[1,2]　渠天轼[1,3]　毕雪林[1,3]　常　鹏[1,3]
（1. 中国建筑股份有限公司埃及分公司，开罗，11865；2. 中建一局集团第三建筑有限公司，北京，100161；3. 中国建筑国际工程公司，北京，100029）

摘　要：以埃及新行政首都CBD项目为例，介绍了一种30.5m超长对拉技术在筏板侧模施工中的应用，结果表明该系统设计简单，施工方便，具有推广意义。通过对其优化设计发现，对于常见的模板系统，由于不同强度材料不合理搭配使用，常会造成高强材料浪费严重。因此为达到设计最优，应合理选择支撑系统各部件的材料。
关键词：筏板施工；超长对拉螺杆；优化设计

Design and application of tie rod supported formwork for raft foundation of CBD project

Hao Jianbing[1,2]　*Qu Tianshi*[1,3]　*Bi Xuelin*[1,3]　*Chang Peng*[1,3]
（1. China State Construction Engrg. Corp. Ltd. (Egypt), Cairo 11865, Egypt;
2. The Third Constr. Co., Ltd. of China Construction First Group, Beijing 100161, China;
3. CSCEC International Operations, Beijing 100029, China）

Abstract: Based on the Central Business District (CBD) project in Egypt, the design and application of a 30.5 meters long tie rod supported formwork for raft foundation is introduced. The results show that the system is easy to design and safe to use. However, the design optimization for the system shows that, for commonly used formwork system, the mix use of different materials with different strength may cause the waste of high strength material. The most optimal design can only be achieved by carefully selecting the appropriate materials during concept design.
Keywords: raft foundation construction; tie rod supported formwork; design optimization

新型边缘抑制型声屏障技术原理及应用

吴赛甲[1]　俞泉瑜[2]　薛嘉鑫[1]　李世航[2]
（1. 天津津铁城市轨道交通工程有限公司，天津，300000；
2. 安境迩（上海）科技有限公司，上海，200030）

摘　要： 随着城市环境噪声污染防治需求的不断增加，考虑了安全及景观美化的直立型声屏障降噪技术方案更具优势，而外形复杂、施工维护困难及存在一定安全隐患的传统顶部降噪装置大大限制了传统声屏障降噪技术方案的发展。边缘抑制型声屏障是基于声屏障边缘空气粒子振动加速导致声音放大的边缘效应而设计，其具有外形简洁、一体化安装、优于传统顶端降噪装置的附加降噪效果等优势，适应性广泛，可为声屏障降噪工程提供更优化的备选方案。

关键词： 边缘效应；声屏障；环境噪声污染；顶端降噪装置

The principle and application of new edge suppression noise barrier technology

Wu Saijia[1]　*Yu Quanyu*[2]　*Xue Jiaxin*[1]　*Li Shihang*[2]
（1. Tianjin Rail Transit Group Co.，Ltd.，Tianjin 300000，China；
2. ANGEL（Shanghai）Technology Co.，Ltd.，Shanghai 200030，China）

Abstract： With the increasing demand of urban environmental noise pollution control technology，the noise reduction technology of vertical noise barrier combined with safety and landscaping has more advantages. Due to the complex shape，construction and maintenance difficulties and safety hazards，the top noise reduction device greatly limits the development of the traditional noise reduction technology. The edge suppression noise barrier is designed based on the edge effect of sound amplification caused by the vibration acceleration of air particles at the edge of the noise barrier. It has the advantages of simple appearance，integrated installation，and superior to the additional noise reduction effect of the traditional top noise reduction device. It has a wide range of adaptability and can provide a more optimized alternative for the noise reduction project of the noise barrier.

Keywords： edge effect；noise barrier；environmental noise pollution；the top noise reduction device

既有运营地铁上方大型深基坑施工工艺研究

权利军　黄　蜀　赵　成

（中建五局第三建设有限公司西北分公司，陕西 西安，710000）

摘　要：本文依托西安市幸福林带建设工程，以风险耦合理论为基础，针对既有运营地铁线路上方深基坑开挖风险进行理论研究及数值分析，建立“先加固、后施工”的工程风险控制理念。对常用的坑内土体加固方式从技术效果、经济效果、施工影响等方面进行评估，提出了“先水泥搅拌桩加固，后分段分层开挖，全过程自动监测”的运营地铁上方深基坑施工技术方案，对基坑进行分层分段开挖，并制定可操作性的施工控制措施。解决了已运营地铁隧道上方深基坑安全施工难题，保障了地铁线路的安全运营。证明利用水泥土搅拌桩对地铁结构施工影响区域进行预加固，遵循分层分块施工原则，是既有运营地铁上方深基坑施工的可靠保障。

关键词：深基坑；邻近地铁；水泥搅拌桩

Above existing operating subway study on the construction technology of large deep foundation pit

Quan Lijun　Huang Shu　Zhao Cheng

(China Construction Fifth Bureau third Construction Co.，Ltd. Northwest Branch，Xi′an，Shanxi 710000，China)

Abstract：This paper based on the construction project of Xingfu forest belt in Xi'an，based on the risk coupling theory，it conducts theoretical research and numerical analysis on the deep foundation pit excavation risk above the existing operation subway line，and establishes the engineering risk control concept of "strengthening first and construction later" . To the common soil reinforcement in the pit from the technical effect，economic effect，construction impact evaluation，put forward the "concrete mixing pile reinforcement，then section layered excavation，the whole process automatic monitoring" operation subway deep foundation pit construction technical scheme，foundation pit layered excavation，and formulate operational construction control measures. It has solved the safe construction problem of the deep foundation pit above the subway tunnel that has been operated，and ensured the safe operation of the subway lines. It is proved that the use of water-soil mixing pile to pre-reinforce the influence area of subway structure construction follows the principle of layered block construction，which is a reliable guarantee for the construction of deep foundation pit above the subway.

Keywords：deep foundation pit；adjacent to the subway；cement mixing pile

莲塘口岸旅检大楼结构舒适度分析与研究

吴荫强[1]　张学民[2]
（1. 深圳市土地投资开发中心，广东 深圳，518031；
2. 深圳华阳国际工程设计股份有限公司，广东 深圳，518038）

摘　要： 莲塘口岸为实现货检和旅检的完全分离，旅检大楼创造性采用立体设计，大型货柜车需从旅检大楼内部通过，对结构舒适度等造成重要影响。通过对类似项目进行楼板振动测试和荷载识别分析，验证出相应荷载可运用至类似频率的结构上，模拟车辆荷载的加载工况。通过建立计算模型，区分不同荷载工况，对货柜车振动效应及楼板舒适度分析，验证了旅检大楼多数区域满足相应标准，对不满足标准区域进行加强措施，并采取减振措施，对类似立体口岸交通在减振隔振方面具有较好的参考价值。
关键词： 车辆荷载；结构舒适度；实测分析；立体口岸交通

Study on application of key construction technology of large angle inclined column in LianTang Port Engineering

Wu Yinqiang[1]　*Zhang Xuemin*[2]
（1. Centre of Shenzhen Land Investment & Development，Shenzhen，Guangdong 518031，China；2. Capol Internationl & Associates Group Shenzhen Branch，Shenzhen，Guangdong 518038，China）

Abstract： In order to realize the complete separation of cargo inspection and passenger inspection at LianTang Port，the passenger inspection building creatively adopts three-dimensional design. Large container truck needs to pass through the interior of the passenger inspection building，which has an important impact on the structural comfort. Through the floor vibration test and load identification analysis of similar projects，it is verified that the corresponding load can be applied to the structure with similar frequency to simulate the loading condition of vehicle load. By establishing the calculation model，to distinguish the different load condition，the container truck vibration effect and comfort analysis of the floor slab，verify the building most areas meet appropriate standards，to strengthen measures，does not meet the standard area and vibration reduction measures，port of similar three-dimensional traffic in terms of vibration isolation has a good reference value.
Keywords： vehicle Load；structural comfort degree；field measurement；three-dimensional port traffic

欠固结软土路基处理施工技术探讨

谢　非

（中国二十冶集团有限公司，上海，201900）

摘　要：本文结合珠海横琴市政道路深厚欠固结淤泥路基处理的难点和特点，BT 项目的实际从设计、施工的角度，分别对复合桩地基处理和真空预压排水固结法处理进行深入的分析和探讨。

关键词：含开山石深厚淤泥；欠固结软土；市政路基；CFG 桩；疏沉控桩；真空预压

Discussion on construction technology of underconsolidated soft soil subgrade treatment

Xie Fei

（China MCC20 Group Co.，Ltd.，Shanghai 201900，China）

Abstract：In the paper，based on the BT project of municipal road in Hengqin，Zhuhai City，considering the difficulties and characteristics of the deep underconsolidated soft subgrade treatment of the project，from the viewpoint of engineering design and construction，the methods of composite piles and vacuum preloading drainage consolidation are analyzed and disscussed separately.

Keywords：deep silt combined rock；underconsolidated soft soil；municipal road；Cement Fly-ash Grave（CFG）pile；control settlement and sparse piles；vacuum preloading

跨一级干线光缆原位保护的复合支护施工技术应用

陈锡华　邢关猛
（中建八局第一建设有限公司，山东 济南，250000）

摘　要：针对紧邻国家一级干线光缆的基坑施工，为了保证光缆和基坑工程的安全，采用一种跨过光缆井道施工高压旋喷桩和双排钢管桩，并采取冠梁连接、锚索外拉和混凝土喷面的复合支护体系进行基坑支护。该支护体系安全可靠，施工便捷，经实践证明支护效果良好，可以在类似工程中应用。

关键词：一级干线光缆；复合支护体系；钢管桩；高压旋喷桩；锚索

Application of composite support construction technology for in-situ protection of optical cables across primary trunk lines

Chen Xihua　Xing Guanmeng
(First Construction Co., Ltd., China Construction 8th Engineering Bureau, Jinan, Shandong 250000, China)

Abstract: For the construction of foundation pits adjacent to the national first-class trunk optical cable, in order to ensure the safety of the optical cable and foundation pit engineering, a high-pressure jet grouting pile and double-row steel pipe piles are used to cross the optical cable shaft, and the crown beam connection and the outer anchor cable are adopted. The composite support system of pulling and concrete spraying surface supports the foundation pit. The supporting system is safe and reliable, and the construction is convenient. Practice has proved that the supporting effect is good, and it can be applied in similar projects.

Keywords: first-class trunk optical cable; composite support system; steel pipe pile; high pressure jet grouting pile; the outer anchor cable

装配式工业厂房塔吊选型及吊装技术

施群凯　杨　前　范作锋
（中国二十冶集团有限公司，上海，201900）

摘　要：建筑工程中，塔吊的选型及布置是决定物料水平和垂直运输快慢的关键，直接影响项目进度。由于塔吊类型众多，塔吊的布置根据现场实际需求不同而不同，尤其在装配式工程中，构件数量多、吊装量大，塔吊的布置和使用更为重要。因此对如何做好塔吊选型布置及与装配式构件安装管理配合是一个难题。本文通过笔者公司一装配式工业厂房项目为例，详细阐述了如何利用BIM等技术解决塔吊的选型，以及装配式构件吊装管理的问题，为其他类似项目提供一些经验借鉴。
关键词：装配式构件；塔吊；BIM技术；吊装

Summary of tower crane lifting application technology for assembly plant

Shi Qunkai　Yang Qian　Fan Zuofeng
(China MCC20 Group Co.，Ltd.，Shanghai 201900，China)

Abstract：In construction projects，the selection and arrangement of tower cranes is the key to determine the speed of horizontal and vertical transportation of materials，which directly affects the project progress. As there are many types of tower cranes，the layout of tower cranes is different according to the actual needs of the site. Especially in assembly-type projects，the number of components is large and the lifting volume is large，so the layout and use of tower cranes are more important. Therefore，it is a difficult problem for how to do a good job of tower crane selection arrangement and cooperation with assembly type component installation management. In this paper，through an assembly-type industrial plant project of our company as an example，we elaborate how to use BIM and other technologies to solve the problems of tower crane selection and assembly component lifting management，so as to provide some experience for other similar projects.
Keywords：prefabricated concrete；tower crane；BIM technology；hoisting

基于等维灰数递补GM（1，1）模型的基坑变形预测研究

张清明[1]　徐　帅[2]　李姝昱[1]

（1. 黄河水利委员会黄河水利科学研究院，河南 郑州，450003；

2. 黄河水资源保护科学研究院，河南 郑州，450003）

摘　要：科学、合理地选择预测模型，及时、准确地反馈基坑变形动态，对基坑安全起着至关重要的作用。灰色GM（1，1）模型预测结果的精度受建模数据列长短的影响，为克服离散、不稳定数据对GM（1，1）模型的不利影响，采用等维灰数递补动态预测法，保持数据列的等维，建立GM（1，1）预测模型，逐个预测依次替补，及时补充和利用新的信息，提高灰区间的白化度，然后利用不同维数GM（1，1）模型构造灰色预测模型群，取其均值作为预测结果，进一步提高预测精度和预测结果的可靠性。应用结果表明，等维灰数递补动态预测GM（1，1）模型预测值的平均相对误差和均方差比值均小于传统GM（1，1）模型的预测值，有效提高了预测精度和预测结果的可靠性。

关键词：基坑；变形；模型；动态预测

Research on prediction of foundation pit deformation based on equal-dimensional grey number progressive GM（1，1）mode

Zhang Qingming[1]　*Xu Shuai*[2]　*Li Shuyu*[1]

（1. Yellow River Institute of Hydraulic Research，YRCC，Zhengzhou，Henan 450003，China；2. Yellow River Water Resources Protection Research Institute，Zhengzhou，Henan 450003，China）

Abstract：Scientific and reasonable selection of prediction model，timely and accurate feedback of foundation pit deformation dynamics，plays a vital role in the safety of foundation pit. The accuracy of grey GM（1，1）model prediction results is affected by the length of the modeling data column. In order to overcome the adverse effects of discrete and unstable data on GM（1，1）model，the GM（1，1）prediction model is established by using the equal-dimensional grey number progressive dynamic prediction method to maintain the equal dimension of the data column，and the new information is supplemented and used in time to improve the whiteness of the grey interval. Then the grey prediction model group is constructed by using different dimensions of GM（1，1）model，and the mean value is taken as the prediction result to further improve the prediction accuracy and reliability of the pre-

diction results. The application results show that the average relative error and mean square deviation ratio of the predicted value of the GM (1, 1) model are less than those of the traditional GM (1, 1) model, which effectively improves the prediction accuracy and reliability of the prediction results.
Keywords: foundation pit; deformation; model; dynamic prediction

钢管混凝土柱缺陷超声波无损检测技术研究

田喜胜[1]　王　军[1]　简宏儒[1]　王登科[1]　黄乐鹏[2]　黄钦全[2]

（1. 中建三局第三建设工程有限责任公司（西南），重庆，400000；

2. 重庆大学，重庆，400044）

摘　要：在高层建筑中钢管混凝土的质量缺陷对建筑结构受力有着不利影响，为保障结构安全，对高层建筑钢管混凝土质量检测是一项必做工作。因钢管混凝土结构的特殊性，在钢管外围采用无损方式检测最为便捷。无损检测作为新型检测方式，其检测效果有待验证。为此，制作了5个厚度各异的钢管混凝土模拟柱，采用超声波对模拟柱内混凝土质量缺陷及缺陷大小进行检测及分析，确定了超声波无损检测技术的可行性。超声波无损检测技术在不破坏钢管混凝土构件的前提下对混凝土缺陷及其大小检测，精度高、可靠性强，可在钢管混凝土柱缺陷检测上推广应用。

关键词：高层建筑；钢管混凝土；超声波；无损检测

Research on ultrasonic nondestructive testing technology of concrete filled Stell tubular column defts

Tian Xisheng[1]　*Wang Jun*[1]　*Jian Hongru*[1]　*Wang Dengke*[1]

Huang Lepeng[2]　*Huang Qinquan*[2]

（1. The Third Construction Engineering Co.，Ltd.（southwest），

Chongqing 400000，China；2. Chongqing University，Chongqing 400044，China）

Abstract：In high-rise buildings，the quality defects of concrete-filled steel tube have different effects on the structural stress of the whole building. Facing the growing number of super high-rise buildings，it has a very broad application prospect to detect the internal defects of steel bar concrete in high-rise buildings. Ultrasonic nondestructive testing technology can detect the size of defects without damaging the concrete-filled steel tubular members，and has high accuracy and reliability. Based on the basic principle of ultrasonic nondestructive testing of concrete-filled steel tube，five concrete-filled steel tube columns with different thickness are made. The size of defects in concrete-filled steel tube is detected and analyzed，which directly reflects the feasibility of ultrasonic nondestructive testing technology.

Keywords：high-rise building；concrete filled steel tube；ultrasonic；nondestructive testing

刚性构造物对沥青路面力学性能影响分析

尹祖超　高　原　邹小龙　张　钰　刘　斌
（长江勘测规划设计研究有限责任公司，湖北 武汉，430010）

摘　要：为了研究刚性构造物对半刚性沥青路面结构受力特性的影响，采用数值模拟手段建立了力学计算模型，分析了刚性墙体位置、格栅设置情况对沥青路面设计指标包括无机结合料稳定层层底拉应力、路基顶压应变、沥青层底竖向压应力的影响。计算结果表明，刚性墙体位于路床范围内时，将改变无机结合料稳定层层底应力情况，使应力变化更复杂；路基顶压应变较正常情况增大约 4 倍；对沥青层底竖向压应力影响不大，均在 0.35～0.4MPa 之间。设计时应将刚性墙体置于路床以下，否则应做破除处理，以利沥青路面结构受力。

关键词：刚性构造物；半刚性沥青路面；受力特性

Influence of rigid structure on mechanic behaviors of asphalt pavement

Yin Zuchao　Gao Yuan　Zou Xiaolong　Zhang Yu　Liu Bin
(Changjiang Survey, Planning, Design and Research Co., Ltd., Wuhan, Hubei 430010, China)

Abstract: In order to study the influence of rigid structure on the mechanical behaviors of semi-rigid asphalt pavement structure, numerical simulation methods were used to establish a mechanical calculation model. The influence of the position of flood wall and geogrid on the design index of asphalt pavement including the tensile stress at the bottom of the inorganic binder stabilized layer, the compressive strain on the top of the roadbed, and the vertical compressive stress at the bottom of the asphalt layer were analyzed. The calculation results show that when the flood wall is located within the range of the road bed, the stress at the bottom of the inorganic binder stabilized layer will be changed, making the stress to change more complicated; the top compressive strain of the roadbed will increase by about 4 times compared with the normal situation, and the vertical compressive stress at the bottom of the asphalt layer will be 0.35-0.4MPa. Thus, the flood wall should be placed below the road bed in the design, otherwise it should be destroyed to facilitate the stress on the asphalt pavement structure.

Keywords: rigid structure; semi-rigid asphalt pavement; mechanic behaviors

纤维加固混凝土梁的受剪性能试验研究

熊吉祥　黄易平　李　杨
（中建五局第三建设有限公司，湖南 长沙，410004）

摘　要：碳纤维增强基复合材料具有良好力学性能，用其加固混凝土梁，测试纤维加固混凝土梁的受剪性能。设计 26 根纤维布加固间隙为 100mm 和 55mm、且粘贴方式不同的试件梁，利用千斤顶和落锤冲击试验机分别设置静载和冲击荷载测试条件，根据不同测试环境下的试件破坏形态，分析试件梁受剪性能。结果表明：静载条件下将纤维布加固间隙设置为 55mm 时，与未加固和加固间隙为 100mm 的两个对照组相比，试件产生的裂痕更短更细分布密度更大，与 5 个差值规范数据相比，受剪承载力差值最大，说明试件梁能更有效抵抗静载压力；冲击荷载条件下采用左右倾斜 45°的方式粘贴纤维布，与垂直粘贴和交叉粘贴纤维布的试件梁相比，试件梁应变时程的能力最强。

关键词：混凝土梁；纤维布；加固间距；冲击荷载；破坏程度；受剪性能

Experimental study on shear behavior of fiber reinforced concrete beams

Xiong Jixiang　Huang Yiping　Li Yang
(3rd Construction Co., Ltd. of China Construction 5th Engineering Bureau, Changsha, Hunan 410004, China)

Abstract: Carbon fiber reinforced matrix composites have good mechanical properties, and they are used to reinforce concrete beams to test the shear performance of fiber-reinforced concrete beams. Design 26 specimen beams with fiber cloth reinforcement gaps of 100mm and 55mm, and different bonding methods, use jacks and drop hammer impact testers to set static load and impact load test conditions respectively, according to the failure modes of specimens under different test environments analyze the shear performance of the specimen beam. The results show that when the fiber cloth reinforcement gap is set to 55mm under static load conditions, compared with the two control groups with unreinforced and reinforced gaps of 100mm, the cracks produced by the specimens are shorter, more finely distributed and denser. Compared with the difference specification data, the difference in shear capacity is the largest, indicating that the specimen beam can resist static load pressure more effectively; under the impact load condition, the fiber cloth is pasted at a 45° angle from left to right, and it is combined with vertical pasting and cross-bonding. Compared with the specimen beam pasted with fiber cloth, the specimen beam has the strongest strain time history capability.

Keywords: concrete beam; fiber cloth; reinforcement spacing; impact load; damage degree; shear performance

钢结构建筑大跨空腹桁架结构设计研究

王田友　潘斯勇　尹洪冰　武云鹏　罗兴隆
（中冶（上海）钢结构科技有限公司，上海，201908）

摘　要：某大体量复杂商业综合体采用框架-中心支撑钢结构体系，对结构体系的确定和采用的大跨度空腹桁架跨街结构布置进行了介绍。考虑施工次序，对跨街空腹桁架结构进行了受力性能分析和防连续倒塌分析；并分析了竖向荷载、地震和温度作用下空腹桁架结构的楼板受力，得出了一些有意义的结论。分析表明所采用的空腹桁架结构能满足受力要求，并采取了相应的结构措施。
关键词：钢结构；框架-支撑结构；空腹桁架；防连续倒塌设计；楼板应力

Research on design of large span vierendeel truss structure of steel structure building

Wang Tianyou　Pan Siyong　Yin Hongbing　Wu Yunpeng　Luo Xinglong
（MCC（Shanghai）Steel Structure Technology Corp.，Ltd.，Shanghai 201908，China）

Abstract：A large-scale complex commercial building complex adopts the frame center braced steel structure system. The determination of the structure system and the large-span Vierendeel Truss cross street structure are introduced. Considering the construction sequence，the mechanical performance and progressive collapse prevention of the vierendeel truss structure are analyzed；The floor stress of Vierendeel truss structure under vertical load，earthquake and temperature is analyzed. The analysis shows that the vierendeel truss structure can meet the stress requirements，and the corresponding structural measures are taken.
Keywords：steel structure；frame-brace structure；vierendeel truss；progressive collapse design；floor stress

装配式空调机房 PP-RP 管材应用研究

王礼杰

（中建五局第三建设有限公司，湖南 长沙，410004）

摘　要： 在建筑暖通系统管道主要以焊接钢管和无缝钢管为主，钢管重量大、隔热性能差、耐腐蚀性低等缺点一直是空调管道的通病。本文提出了采用一种新型且经过改良的复合管，代替传统焊接管道的施工材料及工艺的方式。复合纤维管道由三层组成，内层与外层为 PP-R/PP-RCT 材料，中间层为纤维增强 PP-R/PP-RCT 复合材料。

关键词： 装配式机房；空调管道复合管材；PP-RP 复合管材

Research on application of PP-RP pipe in assembled air conditioning room

Wang Lijie

（The Third Company of China Fifth Engineering Bureau Ltd，
Changsha，Hunan 41004，China）

Abstract：In the building heating and ventilation system，the main pipes are welded steel pipe and seamless steel pipe. The disadvantages of heavy steel pipe，poor heat insulation performance and low corrosion resistance have always been the common faults of air conditioning pipes. A new and improved composite pipe is proposed to replace the traditional welding pipe construction materials and technology. The composite fiber pipe is composed of three layers，the inner and outer layers are made of PP-R/PP-RCT material，and the middle layer is made of fiber reinforced PP-R/PP-RCT composite.

Keywords：prefabricated machine room；air conditioning pipe composite pipe；PP-RP composite pipeterial

城市软弱地基下桩基施工影响分析

张　明　佟安岐　陈长卿　孙北松　魏智锴
（中国建筑工程（香港）有限公司，中国 香港，999077）

摘　要：以香港的一项城市软弱地基情况下的桩基工程为研究对象，通过预钻孔试验、原位标准贯入试验及GCO探针试验等方法，对施工区域进行详尽的探测，并进行理论分析，分析了钻孔灌注桩施工过程中对周围结构及建筑物可能产生的影响，并提出了相应的解决方案，避免了可能产生的沉降等风险。
关键词：桩基施工；城市软弱地基；钻孔灌注桩

Analysis of the influence of bored pile construction under the condition of urban soft foundation

Zhang Ming　Tong Anqi　Chen Changqing　Sun Beisong　Wei Zhikai
(China State Construction Engineering (Hong Kong) Limited, Hong Kong 999077, China)

Abstract: Based on an urban bored pile project in Hong Kong with soft foundations as the research object, the construction area was thoroughly probed through methods such as pre-drilling test, in-situ standard penetration test and GCO probe test. The possible impact on surrounding structures and buildings during the construction of bored piles was analyzed, and corresponding solutions were proposed to reduce the potential risk.
Keywords: pile construction; urban soft foundation; bored pile

大型酒店综合体营业区内无干扰改建施工技术

杨勤禄　邢益江　邓文聪

（中国建筑工程（澳门）有限公司，中国 澳门，999078）

摘　要：为研究大型酒店综合体营业区内无干扰改建施工，依托澳门伦敦人活动中心项目，系统总结了低噪声、低粉尘、无漏水污染的新型绿色施工机械设备工作原理和适用范围，小作业面以及复杂环境的系统性垂直运输方案，以及减少对正常营业干扰施工措施、管理保证措施及安全文明措施要点。同时针对超过 8m 高混凝土墙，选用遥控机器人拆除与碟锯+绳锯切割两种施工方法；针对大面积地台砂浆，选用遥控机器人拆除与碟锯+液压逼爆拆除两种施工方法；分别进行了功效对比、优劣分析。同时针对大面积楼板与大截面次梁拆除，创新设计了以主梁为支撑的平台支架逐仓拆除法，有效避免了拆除风险，提高了拆除效率。该技术为大型混凝土框架结构无干扰拆除和改建施工提供了新的一体化解决方案。

关键词：无干扰施工；营业区内改建工程；吊装运输；机器人拆除

No interference reconstruction technology in the business area of large hotel complex

Yang Qinlu　Xing Yijiang　Deng Wencong

(China State Construction Engineering (Macau) Co., Ltd., Macao 999078, China)

Abstract: In order to study the construction of non-interference reconstruction in the business area of large hotel complex, based on the Londoner event center project in Macau, the working principle and application scope of new green construction machinery and equipment without noise, dust and water leakage pollution, systematic vertical transportation scheme for small working surface and complex environment, and construction measures to reduce normal business interference are systematically summarized Key points of management assurance measures and safety and civilization measures. For concrete wall over eight meters high, two construction methods are selected: Brooke Robot demolition and rope saw + disc saw demolition; For the large-area of floor mortar, two construction methods are selected: Brooke robot demolition and cutting hydraulic burst demolition; The comparison of efficacy and the analysis of their advantages and disadvantages are carried out. At the same time, for the demolition of large area floor and large section secondary beam, the bottom part of main beam is designed as a propping to support the upper part of main beam and removal the upper part of main beam part by part, which effectively avoids the risk of

demolition and improves the removal efficiency. This technology provides a new integrated solution for the construction of large-scale concrete frame structure without interference.
Keywords: no interference construction; reconstruction project; hoisting transportation; cutting and dismantling

基于 InfoWorks ICM 模型的城市排水系统施工影响评估

张　明　佟安岐　陈长卿　魏智锴　孙北松
（中国建筑工程（香港）有限公司，中国 香港，999077）

摘　要：香港中九龙干线启德东工程将在启德河排水口建造临时工程，施工过程中将减小排洪河道过流断面。通过 InfoWorks ICM 建立城市排水水力模型，模拟施工过程中对启德河、箱型暗渠，以及上游排水系统的影响，根据模拟结果进行影响评估并提出建议。
关键词：排水系统；影响评估；水力模型；InfoWorks ICM

Impact assessment of construction on stormwater system based on InfoWorks ICM model

Zhang Ming　Tong Anqi　Chen Changqing　Wei Zhikai　Sun Beisong
(China State Construction Engineering (Hong Kong) Limited, Hong Kong 999077, China)

Abstract: Temporary works will be constructed at the Kai Tak River outlet for the Kai Tak East project of the Central Kowloon Route in Hong Kong. During the construction process, the overflow section of the flood discharge channel will be reduced. Through the simulation of InfoWorks ICM, the urban drainage hydraulic model was established to simulate the impact of the construction on the drainage system of Kai Tak River, box culvert and the upstream branch drains, and the impact assessment was made based on the simulation results and suggestions were put forward.
Keywords: storm drainage system; effects evaluation; hydraulic model; InfoWorks ICM

混凝土结合面处理工艺及检测方法研究现状

姜国永　张德利　张新江　黄选明　黄广华　贺　鹏
（中国建筑科学研究院有限公司，北京，100013）

摘　要： 新旧混凝土结合面往往是构件的薄弱部位，其粗糙度对构件的承载力和刚度有很大的影响，对混凝土结合面进行粗糙度处理是提高结合面整体性能的关键。合适的处理方法，准确的检测技术，科学的评价体系以及定量的验收标准成为工程界的亟需。本文总结了常见的粗糙处理方法及检测方法，其中基于 3D 激光扫描法是一种非接触式的无损检测方法，在实际工程中有广阔的应用前景。目前规范标准还没有给出比较明确的量化指标，以机械行业和岩石力学行业的表述方式作为参考，探讨混凝土领域中粗糙度的量化评价方法。

关键词： 粗糙度；处理方法；检测技术；评价方法

Research status of concrete joint surface treatment technology and detection method

Jiang Guoyong　Zhang Deli　Zhang Xinjiang　Huang Xuanming
Huang Guanghua　He Peng
(China Academy of Building Research Co., Ltd., Beijing 100013, China)

Abstract: The bonding surface of new and old concrete is often the weak part of the component, and its roughness has a great influence on the bearing capacity and rigidity of the component. The roughness treatment of the concrete bonding surface is the key to improving the overall performance of the bonding surface. Appropriate processing methods, accurate detection technology, scientific evaluation system and quantitative acceptance standards have become urgent needs of the engineering community. This article summarizes the common rough processing methods and detection methods. Among them, the 3D laser scanning method is a non-contact non-destructive detection method, which has broad application prospects in practical engineering. At present, the norms and standards have not given clear quantitative indicators. Taking the expression methods of the machinery industry and the rock mechanics industry as a reference, we will discuss the quantitative evaluation method of roughness in the concrete field.

Keywords: roughness; processing method; detection technology; evaluation method

贝雷方柱支撑体系在大跨度钢结构中的应用

蒋　卫　张位清　戴超虎　廖　飞　邓正磊
（中建五局第三建设有限公司，湖南 长沙，410004）

摘　要：针对大跨度钢结构高空原位散装法的施工特点，以长沙黄花综合保税区进出口商品展示交易中心钢结构项目为实例，对比了各种常用临时支撑体系选用时的优劣势，介绍了贝雷方柱支撑体系的高度选取原理与细部连接设计，采用有限元软件对贝雷方柱支撑体系正常使用状态和极限偏心受压状态时的承重能力仿真分析，确保了其强度、变形及稳定性满足要求。并对贝雷方柱支撑体系的安装和拆卸做了阐述。最后总结了贝雷方柱支撑体系应用时的实践经验，为后续类似工程提供了实例借鉴。

关键词：贝雷方柱；贝雷片；支撑胎架；支撑体系；大跨度钢结构

Application of Bailey square column support system in large span steel structure

Jiang Wei　Zhang Weiqing　Dai Chaohu　Liao Fei　Deng Zhenglei
(The Third Construction Co., Ltd. of China Construction Fifth Engineering Bureau, Changsha Hunan 410004, China)

Abstract: At present, large-span steel structure is more and more widely used in public buildings. The main construction methods of large-span steel structure are bulk, lifting, jacking, sliding, etc. Compared with lifting, jacking, sliding and other construction methods, bulk method has the advantages of small risk, low cost, small difficulty and so on. Therefore, bulk method is widely used in large-span steel structure. However, in bulk method, the selection and design of support system is the key. Taking a steel structure project of a exhibition center in Changsha as an example, this paper introduces the selection and design of Bailey square column support system in detail, uses the finite element software to simulate and analyze the support system, and expounds the installation and disassembly of Bailey square column, which provides experience summary for similar projects in the future.

Keywords: Bailey square column; Bailey plate; support jig; support system; large span steel structure

构造柱钢筋模块化绿色施工技术研究与拓展

杨 明 黄日欢 刘家龙
（广西建工集团第五建筑工程有限责任公司，广西 柳州，545001）

摘 要：分析现有构造柱钢筋施工碳排放高的原因，提出一种局部模块化，工具化的PVC套模施工工艺，然后采用侧翼包裹、加设锥形套筒、采取180°上部端头等形式对该施工工艺进行优化，并从选材、定位、加固及拆模等方面阐述其施工控制要点，最后对其进行应用拓展，并就其推广价值进行总结。
关键词：构造柱钢筋；绿色施工；套模；模块化；碳排放

Research and development of modular green construction technology of structural column reinforcement

Yang Ming Huang Rihuan Liu Jialong
（Guangxi Construction Engineering Group No. 5 Construction Engineering Co.，Ltd.，
Liuzhou，Guangxi 545001，China）

Abstract：Analyzed the reasons for the high carbon emission of the steel reinforcement construction of the existing structural columns，and proposed a partially modularized and tooled PVC sleeve mold construction technology，and then used the side wing wrapping，additional tapered sleeves，and 180°upper end. The construction method is optimized，and its construction control points are explained from the aspects of material selection，positioning，reinforcement and mold removal，and finally its application is expanded，and its promotion value is summarized.
Keywords：structural column reinforcement；green construction；nesting；modularization；carbon emission

小直径芯样折断力与普通混凝土力学性能指标关系试验研究

罗居刚[1,2]　郜洪生[1,2]　黄从斌[1,2]　张今阳[1,2]

（1. 安徽省（水利部淮河水利委员会）水利科学研究院，安徽 合肥，230088；

2. 安徽省建筑工程质量监督检测站，安徽 合肥，230088）

摘　要：本文探讨了直径55mm芯样抗折法推定混凝土强度的技术原理，通过制作144组432个混凝土芯样试件，并与对应数量的标准试件进行对比。测得不同龄期、不同强度下芯样的折断力与标准试件下的抗压强度、抗折强度及劈拉强度。同时，分别以标准试块的抗压强度、劈裂抗拉强度和抗折强度为因变量，采用最小二乘法对试验数据进行拟合，分别给出芯样折断力与混凝土抗压强度、劈裂抗拉强度、抗折强度关系曲线，得到了拟合方程，研究结果可供工程结构混凝土强度检测与理论研究参考。

关键词：微小芯样；折断力；普通混凝土力学性能；相关关系；试验研究

装配式 L 型 UHPC 楼梯板受弯性能试验研究

李新星　周　泉　李水生
（中国建筑第五工程局有限公司，湖南 长沙，410000）

摘　要： 结合当前国内外装配式构件轻量化的发展趋势，提出一种装配式 L 型 UHPC 楼梯板。研究了配筋和钢纤维对楼梯板抗弯性能和破坏形式的影响，结果表明：TB-2 未加入钢纤维，破坏为脆性破坏，楼梯板最大挠度均发生在梯板跨中边缘处，钢纤维长径比和钢筋对试件承载能力均有影响，且钢纤维对楼梯板承载力作用大于钢筋，钢纤维长径比值越大，试件延性越强；TB-3、TB-4 采用 UHPC 浇筑且配置钢筋，钢纤维对裂缝的两边混凝土有桥联锚固作用，延缓了裂缝的开展，在正常使用阶段，楼梯板钢筋均未屈服，且试验得到的承载力远大于按荷载标准组合的荷载值，裂缝宽度均小于 0.1mm。
关键词： UHPC；L 型楼梯板；挠度；承载力

Experimental study on bending performance of assembly L-type UHPC stair plate

Li Xinxing　Zhou Quan　Li Shuisheng
(China Construction Fifth Engineering Division Co., Ltd., Changsha, Hunan 410000, China)

Abstract: Combined with the development trend of light weight fabricated components at home and abroad, a fabricated L-type UHPC stair slab is proposed. The influence of reinforcement and steel fiber on the flexural performance and failure mode of stair slab is studied. The results show that the failure of TB-2 without steel fiber is brittle failure, and the maximum deflection of stair slab occurs at the mid span edge of stair slab. The ratio of length to diameter of steel fiber and steel bar have influence on the bearing capacity of specimen, and the effect of steel fiber on the bearing capacity of stair slab is greater than that of steel bar, The ductility of the specimen is stronger; TB-3 and TB-4 are poured with UHPC and reinforced. Steel fiber can bridge and anchor the concrete on both sides of the crack, which delays the development of the crack. In the normal use stage, the reinforcement of the stair slab does not yield, and the bearing capacity obtained from the test is far greater than the load value according to the load standard combination, and the crack width is less than 0.1mm.
Keywords: ultra high performance concrete; L-shaped stair slab; deflection; bearing capacity

临近地铁超长异性地下室外墙降噪防开裂技术优化方法

彭　阳　王朝辉　蒋雨明
（中国建筑第五工程局有限公司，湖南 长沙，410000）

摘　要： 常规优化方法通过计算外墙弹性模量、墙体应力场分析外墙开裂机理，分析结果偏差较大，导致外墙受到临地铁深基坑工程的作用力，产生较高的温度应力和剪应力，为此，提出临近地铁超长异性地下室外墙降噪防开裂技术优化方法。采用仿真软件，定义外墙材料本构关系，设置模型荷载及边界，建立临地铁深基坑工程地下室外墙仿真模型；计算地下室外墙混凝土的抗压强度等参数值，分析临地铁深基坑工程地下室外墙开裂机理；依据分析结果，考虑临地铁深基坑工程，在地下室外墙上施加的荷载，增加一附加力偶，在外墙上产生附加弯矩，抵消临地铁深基坑工程施加的力，实现临近地铁超长异性地下室外墙降噪防开裂技术优化。实验结果，研究方法较两组常规方法，产生的温度应力分别小0.4929MPa、0.4246MPa，产生的剪应力分别小0.35MPa、0.28MPa，具有较优的外墙防开裂优化性能。

关键词： 地铁深基坑；深基坑工程；地下室；外墙；防开裂；技术优化

Optimization method of anti cracking technology for basement exterior wall of deep foundation pit engineering near subway

Peng Yang　Wang Zhaohui　Jiang Yuming
(China Construction Fifth Engineering Division Co., Ltd., Changsha, Hunan 410000, China)

Abstract: The conventional optimization method analyzes the cracking mechanism of the external wall by calculating the elastic modulus of the external wall and the stress field of the wall. The deviation of the analysis results is large, which causes the external wall to be subjected to the force of the subway deep foundation pit engineering, resulting in higher temperature stress and shear stress. Therefore, the optimization method of anti cracking technology for the basement external wall of the subway deep foundation pit engineering is proposed. The simulation software is used to define the constitutive relationship of exterior wall materials, set the model load and boundary, and establish the simulation model of basement exterior wall of deep foundation pit engineering near subway; Based on the calculation of the compressive strength and other parameters of the concrete of the basement exterior wall, the cracking mechanism of the basement exterior wall of the deep foundation pit project is analyzed; According to the analysis results, considering the subway deep

foundation pit engineering, an additional couple is added to the load applied on the outdoor wall of the basement, which produces additional bending moment on the external wall, offsets the force applied by the subway deep foundation pit engineering, and realizes the anti cracking technology optimization of the basement external wall of the subway deep foundation pit engineering. The experimental results show that the temperature stress produced by the research method is 0.4929MPa and 0.4246MPa lower than that produced by the two groups of conventional methods, and the shear stress produced by the research method is 0.35MPa and 0.28MPa lower than that produced by the two groups of conventional methods, respectively.
Keywords: subway deep foundation pit; deep foundation pit engineering; basement; external wall; anti cracking; technology optimization

大型铝架设计及稳定性研究

林东海　符鸣晓
（中国建筑工程（香港）有限公司，中国 香港，999077）

摘　要：为研究工程维修时搭建的户外大型铝架搭建时对原有结构及地基的影响，以及铝架自身在强风荷载及使用过程中的振动、位移，进行结构监测系统的安装，在铝架长时间的使用过程中，随时监控结构变形及铝架位移情况。根据现有规范对临时结构的要求，比对现场数据与设计值的要求。根据目前铝架的结构设计以及现场的搭建情况，数据表明此类铝架具有良好的刚度及强度，连接构造安全可靠，满足现场使用要求以及现有的规范设计要求，可以大量在维修工程或新建工程中作为外墙工作平台应用。
关键词：铝架；工作平台；维修工程；结构监测；振动监测；位移监测

Study on design and stability of aluminum scaffolding

Lin Donghai　Fu Mingxiao
(China State Construction Engineering (Hong Kong) Limited, Hong Kong 999077, China)

Abstract: In order to study the impact of the large outdoor aluminum scaffolding as a working platform during maintenance works, as well as the vibration and displacement of the scaffolding itself under strong wind load, the structural monitoring system is introduced. During the use progress, the structural elements deformation, foundation settlement and the displacement of scaffolding is monitored time to time. According to the current used code of practice requirements, comparison between design value and actual value is compared and found satisfactory. Based on the structural design of scaffolding and the one built on site, the monitoring records shows this kind of scaffolding has good rigidity and strength, the connection is safe and reliable and can totally fulfill the on-site use and also the design requirements. Thus it can be widely used in future external wall maintenance works or even new building works as a working platform.
Keywords: aluminum scaffolding; working platform; maintenance works; structural monitoring; vibration monitoring; displacement monitoring

沈阳中海地产住宅项目的绿色实践

王　晋　李东岳　孙麟博　张　志
（中海地产（沈阳）有限公司，辽宁 沈阳，110000）

摘　要： 以中海润山府项目为例，从绿色评价体系、重点绿色技术应用、设计全过程绿色建筑指标控制等方面进行绿色建筑的全生命周期的管控，最终形成低碳、健康、舒适、节能可持续的绿色成果。
关键词： 绿色建筑；碳中和

Green practice of Shenyang Zhonghai Estate Residential Project

Wang Jin　Li Dongyue　Sun Linbo　Zhang Zhi
(China Overseas Development Group Co., Ltd., ShenYang, Liaoning 110000, China)

Abstract: Take Zhonghai Runshan House Project as an example, the whole life cycle control of green building is carried out from the aspects of green evaluation system, the application of key green technology, and the whole green building index control of the design process. Finally, low carbon, healthy, comfortable, energy saving and sustainable green results.
Keywords: green building; carbon-neutral

单塔吊基础扩建形成双塔吊联合承台基础设计分析

唐宇轩[1]　周　泉[1,2]

（1. 中国建筑第五工程局有限公司，湖南 长沙，410004；2. 湖南大学，湖南 长沙，410082；）

摘　要：武汉华侨城杨春湖商务区 A 地块塔楼总高度 275m，塔楼地下室 4 层，因施工组织实施方案变化在原有构造基础上进行设计变更，现需在原塔吊承台基础外侧进行改扩建，形成双塔吊共用联合承台的施工方案。结合塔吊设备技术参数、施工方案及双塔吊共用联合承台的设计方案，应用有限元分析软件 ABAQUS 建立双塔吊共用联合承台的实体模型，进行塔吊在多种工况荷载作用下的受力模拟分析和结构设计方案复核计算。结果表明承台基础的混凝土未出现破坏损伤，钢筋未出现屈服。此外，计算过程中对作用荷载和混凝土强度等级都赋予了一定比例的安全系数，塔吊基础钢架（钢柱）的埋置加强措施也被简化，因此相应计算分析具有一定的结构安全储备。

关键词：改扩建设计；ABAQUS；混凝土损伤塑性模型；有限元；力学性能

Design analysis of double tower crane combined bearing platform foundation formed by single tower crane foundation expansion

Tang Yuxuan[1]　*Zhou Quan*[1,2]

（1. China Construction Fifth Engineering Division Co. , Ltd. , Changsha, Hunan 410004, China; 2. Hunan University, Changsha, Hunan 410004, China）

Abstract: The total height of the tower in block A of Yangchunhu Business District of Wuhan overseas Chinese city is 275 meters, and the basement of the tower is 4 floors. Due to the change of the construction organization and implementation plan, the design is changed on the basis of the original structure. Now it is necessary to reconstruct and expand the outside of the original tower crane bearing platform foundation, forming the construction plan of double tower cranes sharing the combined bearing platform. Combined with the technical parameters of tower crane equipment, construction scheme and design scheme of double tower crane sharing the combined bearing platform, the solid model of double tower crane sharing the combined bearing platform is established by using the FEA software ABAQUS, and the stress simulation analysis and structural design scheme review calculation of tower crane under various load conditions are carried out. The results show that the concrete of pile cap foundation is not damaged, and the steel bar is not yielding. In addi-

tion, in the calculation process, a certain proportion of safety factor is given to the action load and concrete strength grade, and the embedded strengthening measures of tower crane foundation steel frame (steel column) are also simplified. Therefore, the corresponding calculation and analysis has a certain structural safety reserve.
Keywords: extension design; ABAQUS; damaged plasticity model for concrete; finite element; mechanical property

可周转式组合型声测管施工技术

祁明军　许　鹏　张海波　杨　柳　程　杭

摘　要： 大型钻孔灌注桩属隐蔽工程，目前桩基普遍利用预埋声测管进行超声检测，但全国各大设计院采用声测管材质、厚度各不相同并且多数工程均使用一次性声测管，试验后将声测管留在桩基内造成材料浪费，且受声测管材料或施工过程影响，无法进行超声波检测的情况时有发生。结合项目桩基组合型声测管技术，介绍了一种可周转式组合型声测管施工技术，并对该施工技术的工艺流程及应用效果进行了简单讲解，可为其他同类项目的施工提供参考。

关键词： 可周转；组合型声测管；桩基检测

The utility model relates to a construction technology for a turnable combined acoustic measuring tube

Qi Mingjun　Xu Peng　Zhang Haibo　Yang Liu　Cheng Hang

Abstract: Large bored piles belong to take cover engineering, pile foundation currently widespread use of embedded tube in ultrasonic testing, but the major national design institute using acoustic detecting tube material, the thickness of each are not identical and should be used in most engineering disposable tube, after test will be waste material of acoustic detection tube in pile foundation, and influenced by sound tube material or construction process, It is not uncommon for ultrasound tests to be performed. Combined with the construction technology of pile foundation combined acoustic measuring tube, this paper introduces a construction technology of revolving combined acoustic measuring tube, and briefly explains the technological process and application effect of the construction technology, which can provide reference for other similar projects.

Keywords: can-turnaround; combined type acoustic measuring tube; pile foundation detection

高层建筑铝合金模板体系中全现浇外墙质量控制研究

牟　钰　孙久威
（中国建筑第五工程局有限公司，辽宁 沈阳，110000）

摘　要： 近年来，全现浇外墙因具有施工速度快、绿色环保、外墙一次成型，减少二次施工及防止外墙渗漏等优点而在工程中广泛应用，但现在存在设计单位，对现浇混凝土墙体替代砌筑墙体引起结构受力状态、刚度改变等问题，在设计时未充分考虑；施工单位对现浇外墙施工不规范，现浇外墙相关验收环节缺失等问题；质量监督机构对全现浇外墙质量监管不到位等问题。本文从质量监管角度，对全现浇外墙的质量监管提出一些措施，督查各方责任主体对全现浇外墙的质量给予重视。

关键词： 铝合金模板；全现浇外墙；质量控制；结构性能

嘉兴文化艺术中心大悬挑开合屋盖计算分析

黄银春　潘斯勇　王田友　罗兴隆
（中冶（上海）钢结构科技有限公司，上海，201908）

摘　要：本文对嘉兴文化艺术中心开合屋盖的设计思路和荷载取值做了详细介绍，开合屋盖分析需要考虑基本状态、非基本状态和运行状态三种状态，每种状态荷载取值有所不同。以其中一个馆开合屋盖作为典型，对其单独分析和协同分析结果做了比较。取其计算结果的包络内力，对九管汇交的复杂节点和台车附近内力较大的节点进行有限元分析，并对整个活动屋盖进行稳定性分析。
关键词：开合屋盖；单独分析；协同分析；有限元分析；稳定性分析

Calculation and analysis of the cantilevered retractable roof of Jiaxing Culture and Art Centre

Huang Yinchun　Pan Siyong　Wang Tianyou　Luo Xinglong
（MCC Shanghai Steel Structure Technology Co., Ltd., Shanghai 201908, China）

Abstract: In this paper, the design idea and load value of retractable roof of Jiaxing Culture and Art Centre were introduced in detail. Analysis of retractable roof needs to consider three states: basic state, non-basic state and operating state. The load value of each state was some different. The results of individual analysis and collaborative analysis are compared by taking one of retractable roofs for example. The finite element analysis was carried by taking the envelope internal force of the calculated results for complex nine pipes joints and joint near the trolley. Also, the stability analysis was carried out for the whole retractable roof structure.
Keywords: retractable roof; individual analysis; collaborative analysis; finite element analysis; stability analysis

泥质粉砂层群井降水选控及技术分析研究

宋绪旺　王五洋　肖　超　罗桂军　周雄威
（中建五局土木工程有限公司，湖南 长沙，410004）

摘　要：针对南宁市轨道交通4号线一期工程施工总承包01标土建1工区那历村站—那洪立交站区间联络通道开挖泥质粉砂层出现高压喷水、喷砂的问题，对那历村站—那洪立交站区间联络通道地质条件以及降水过程进行了详细的技术分析，探讨了联络通道出现高压喷水、喷砂的原因，在此基础上提出了联络通道群井降水加两台阶留核心土法开挖相结合的处治对策并付诸应用。实践表明该区间联络通道处治对策是合理有效的，对广西壮族自治区及类似地质区地铁区间联络通道开挖降水处治具有很好的参考价值。
关键词：地铁区间；泥质粉砂层；群井降水；技术分析研究

Selected control and technical analysis of well dewatering in silty sand formation group

Song Xuwang　Wang Wuyang　Xiao Chao　Luo Guijun　Zhou Xiongwei
（CCFED Civil Engineering Co.，Ltd.，Changsha，Hunan 410004，China）

Abstract：In view of the problem of high-pressure water and sand jetting occurred in the excavation of the silty sand layer of the connecting passage between the Nanlicun station and Nahong interchange station in the 1 St Phase Project of Nanning Rail Transit Line 4，in this paper，the geological conditions and precipitation process of the connecting passage in na-na interval are analyzed in detail，and the causes of high-pressure water and sand spraying in the connecting passage are discussed，on the basis of this，the paper puts forward the treatment countermeasure of the combination of the connection channel group well dewatering and the two-step retaining core soil excavation method. The practice shows that the treatment measures are reasonable and effective，which has a good reference value for the excavation and precipitation treatment of the Guangxi Zhuang Autonomous Region and similar geological areas.
Keywords：metro section；argillaceous silt layer；group well precipitation；technical analysis study

园林景观项目应用海绵城市的施工质量控制要点浅析

李国鸿

（广州建筑工程监理有限公司，广东 广州，510030）

摘　要：目前阶段，园林景观建设过程中投入了建设海绵城市的相关理念，海绵城市是一种分散式的雨水管理模式，它强调雨水资源在城市中间的渗透、过滤、净化、储存、蓄流和缓慢排入城市市政管网、河道的过程，它能控制雨水径流、促进水的循环利用，有效地解决城市积水、内涝等问题。但是目前海绵城市的工程质量管理存在较多的不足，涉及是多方面的，规范的不完善、施工无经验或经验不足、施工顺序安排不合理、设计或深化设计的不成熟、有关海绵城市方面的质量检查标准不健全等因素都会直接或间接影响海绵城市质量管理。

关键词：园林景观；应用；海绵城市；质量控制

Analysis on the construction quality control points of sponge city in landscape projects

Li Guohong

(Guangzhou Construction Engineering Supervision Company, Guangzhou, Guangdong 210030, China)

Abstract: At present, sponge city is a decentralized rainwater management mode, which emphasizes the process of rainwater resources penetrating, filtering, purifying, storing and storing in the middle of the city and slowly discharging into the municipal pipe network and river. It can control rainwater runoff, promote water recycling, and effectively solve the problems of urban water accumulation and waterlogging. However, at present, there are many deficiencies in the engineering quality management of sponge city, involving various factors, such as imperfect specifications, inexperienced or inexperienced construction, unreasonable construction sequence arrangement, immature design or deepening design, and unsound quality inspection standards related to sponge city, which will directly or indirectly affect the quality management of sponge city.

Keywords: landscape architecture; application; sponge city; quality control

浅析预制装配式结构的节点连接技术

张　陆
（中建科技集团有限公司北京分公司，北京，100176）

摘　要：在我国新技术以及相关装配式政策的推动下，建筑业的深化改革和转型升级已经是未来的发展趋势，为了提升我国建筑行业的核心竞争力，顺应时代潮流的绿色节能环保的预制装配式结构建筑应运而生。目前，在我国装配式结构的建筑行业领域中节点的连接方式是其重要组成部分。装配式结构建筑中的节点连接方式的选用，会对预制装配式建筑的整体结构安全至关重要，所以在施工过程中需要对各类型的节点连接方式进行择优选用最经济合理的连接方式。

关键词：预制装配式结构；建筑节点；连接方式

Analysis on joint connection technology of prefabricated structure

Zhang Lu
(Beijing Branch of China Construction Technology Group Co.，Ltd.，Beijng 100176，China)

Abstract：under the promotion of China′s new technology and related prefabricated policies，the deepening reform，transformation and upgrading of the construction industry has become the future development trend. In order to enhance the core competitiveness of China′s construction industry，green energy-saving and environmental protection prefabricated structure buildings that conform to the trend of the times emerge as the times require. At present，in the field of prefabricated structure construction industry in China，the connection mode of nodes is an important part. The selection of joint connection mode in prefabricated structure building is very important to the overall structural safety of prefabricated building，so in the construction process，it is necessary to select the best joint connection mode of various types，and select the most economical and reasonable connection mode.

Keywords：prefabricated structure；building nodes；connection mode

装配式建筑塔吊选型及布置的应用与研究

郭 瑾[1] 姜福明[2]
（1. 中建科技集团有限公司北京分公司，北京，100176；
2. 青岛公共住房建设投资有限公司，山东 青岛，266071）

摘 要： 结合目前正在参与施工建设的北京市延庆新城03街区21、25地块项目，探究如何通过优化塔吊选型、布置方法创新以及吊具的选用等，以满足现场装配式构件对于塔吊相关吊重、吊次等要求的情况下优化作业时间减少施工成本。本文主要结合实际工程应用，对装配式建筑塔吊的选型及布置进行详细的分析，并提出具体相关要求。
关键词： 装配式建筑；群塔；塔吊

Model selection and layout of prefabricated building tower crane

Guo Jin[1] *Jiang Fuming*[2]
（1. Beijing Branch of China Construction Technology Group Co.，Ltd.，Beijng 100176，China；2. Qingdao Public Housing Construction Investment Co.，Ltd.，Qingdao，Shandong 266071，China）

Abstract： Combining with the currently participating in the construction and construction of the blocks 21 and 25 in Block 03，Yanqing New City，Beijing，we will explore how to optimize the selection of tower cranes，the innovation of layout methods，and the selection of spreaders to meet the requirements of the site-assembled components for the related lifting，Optimize operation time and reduce construction costs in the case of hoisting requirements. This article mainly combines actual engineering applications to analyze the selection and layout of prefabricated building cranes in detail，and puts forward specific related requirements.
Keywords： prefabricated building；tower group；tower crane

装配式建筑坐浆法施工的研究

李　木[1]　黄金冶[1]　姜福明[2]　曹龙伟[1]　王英杰[1]
（1. 中建科技集团有限公司北京分公司，北京，100000；
2. 青岛公共住房建设投资有限公司，山东 青岛，266071）

摘　要：为研究装配式建筑中灌浆施工如何一次合格，在墙板吊装中采用坐浆法施工进行反复试验。试件为装配式转换层外墙、标准层外墙、伸缩缝内墙、内墙等构件。详细考察施工全过程和灌浆结束后孔道情况，分析灌浆法、坐浆法对比分析。根据现有规范评价。试验研究表明，坐浆法施工对提高灌浆一次合格率更有保障，相对于灌浆法更多避免后期补浆等不利后果，坐浆法施工更适合装配式建筑施工，对于提高灌浆一次合格率更为可靠。

关键词：坐浆法施工；提高灌浆一次合格率

Study on the construction of prefabricated building with grouting method

Li Mu[1]　*Huang Jinye*[1]　*Jiang Fuming*[2]　*Cao Longwei*[1]　*Wang Yingjie*[1]
(1. China Construction Science & Technology Group Co., Ltd., Beijing 100000, China;
2. Qingdao Public Housing Construction Investment Co., Ltd., Qingdao, Shandong 266071, China)

Abstract: In order to study how to qualified grouting construction in prefabricated building, the block grouting method was used in wall panel hoisting and repeated tests were carried out. The specimens are assembled external wall of conversion layer, external wall of standard layer, internal wall of expansion joint, internal wall and other components. Investigate the whole process of construction and the tunnel situation after grouting in detail, and analyze the grouting method and grouting method. Evaluate against existing specifications. The experimental study shows that the construction of grouting method is more guaranteed to improve the primary qualified rate of grouting. Compared with the grouting method, the construction of grouting method is more suitable for prefabricated building construction and more reliable to improve the primary qualified rate of grouting.

Keywords: block slurry method construction; improve the primary qualified rate of grouting

专题四　数　字　建　造

大型建筑企业 BIM 体系建设经验研究

田　华　稂洪波　赵庆祥　牛孜飏
（中国建筑第五工程局有限公司，湖南 长沙，410007）

摘　要：本文对中建五局自 2008 年以来在企业级 BIM 体系建设方面的实践经验进行总结，将企业 BIM 建设体系分解为目标体系、组织体系、制度体系、资源体系、考核体系、推广体系和研发体系。明确企业 BIM 应用原则、项目层面应用目标和组织层面应用目标，引领企业 BIM 建设和项目 BIM 应用。搭建从局至项目部的三级组织架构，制定配套的 BIM 管理制度。以超过 1400 人的人才梯队为后盾，以 BIM 软件和构件库资源为工具，提升企业项目 BIM 应用能力。通过标杆项目的示范引领，带动全局项目 BIM 应用。通过对二级单位和项目部 BIM 应用进行考核，促进二级单位 BIM 体系建设和项目 BIM 应用。通过 BIM 工具和平台软件研发，解决现有 BIM 软件的不足，提升 BIM 工作效率。中建五局十余年的实践结果表明，这套企业级 BIM 建设体系行之有效，可供其他类似企业借鉴。
关键词：BIM；企业管理；体系建设；新业务管理

Enterprise-level BIM systems for large-scale construction firms: an empirical study

Tian Hua　Lang Hongbo　Zhao Qingxiang　Niu Ziyang
（China Construction Fifth Engineering Bureau Corporation Ltd.，Changsha，
Hunan 410007，China）

Abstract：The experience of enterprise-level BIM development from China Construction Fifth Engineering Bureau Corporation (CSCEC5B) was concluded and the enterprise-level BIM development systems was decomposed into objective systems，hierarchy systems，regulation systems，resource systems，evaluation systems，standardization systems and R&D systems. First，the principles of enterprise BIM application were clarified，project-level application goals and organization-level application goals were established to guide enterprise-level BIM development and project-level BIM implementation. Then，a three-level BIM management and implementation hierarchy from the bureau down to construction project is established and BIM standards and regulations were formulated to manage BIM implementation. Over 3000 engineers were trained and a BIM team with more than 1400 BIM professionals was formed to support BIM execution. Several BIM authoring software and a component library formed the BIM resource pool to accelerate BIM implementation. In the standardization system，excellence projects were put as benchmarking to lead other projects

in BIM execution. Finally, a R&D system was set-up to solve drawback of current software and improve BIM tasks efficiency. The viability of this enterprise-level BIM development methodology has been demonstrated by practice from CSCEC5B over last 10 years and can be learned by similar organizations.
Keywords: BIM; corporation management; systems establishment; new business management

数字建造在鄂州花湖机场的施工应用

吴 军[1] 路 浩[2] 张 磊[1] 王立红[1] 李东明[1]

(1. 中国建筑第八工程局有限公司中南分公司，湖北 武汉，430000；

2. 中国建筑第八工程局有限公司西南分公司，四川 成都，610000)

摘 要：本文以鄂州花湖机场的施工管理特点，重点阐述施工阶段项目管理的数字化应用。在数字化施工应用中，采用前沿的信息化手段，结合物联网、BIM技术和施工管理平台的集成应用，将涉及项目管理的人、机、料、法、环等施工要素和质量、造价进行深入串联，旨在为项目的管理优化和提质增效进行创新，同时也为我国民航建设施工数字化转型而探索方向。

关键词：数字化；信息化；物联网；BIM技术；施工要素；数字化平台；数字化转型

Construction application of digital construction at Ezhou Huahu Airport

Wu Jun[1] *Lu Hao*[2] *Zhang Lei*[1] *Wang Lihong*[1] *Li Dongming*[1]

(1. China State Construction Eighth Engineering Bureau Co. , Ltd. ,

Zhongnan Branch, Wuhan, Hubei 430000, China;

2. China State Construction Eighth Engineering Bureau Co. , Ltd. , Xinan Branch,

Chengdu, Sichuan 610000, China)

Abstract: This article focuses on the digital application of project management in the construction phase based on the construction management characteristics of Ezhou Huahu Airpot. In the application of digital construction, the use of cutting-edge information methods, combined with the integrated application of the Internet of things, BIM technology and construction management platform, involve the construction elements, such as man, machine, material, method, and environment, as well as elements such as quality and cost of project management. Such methods can help facilitate innovation, project management optimization, quality and efficiency improvement, and at the same time explore the direction for the digital transformation of civil aviation construction in China.

Keywords: digitalization; informationalization; the Internet of Things; BIM technology; construction elements; improving the quality and increasing the efficiency; digital transformation

大跨度大截面变曲率跨层桁架施工精度控制措施及方法

董 涛 张文斌 张玉宽 马晓伟 贺 斌
（中国建筑第八工程局钢结构工程公司，上海，200120）

摘 要：随着我国社会和经济的飞速发展，个性化、别具风格、有意蕴的钢结构建筑不断涌现，大跨度大截面变曲率跨层桁架钢结构因其造型新颖、结构强度稳定性好等特点被应用到建筑结构当中，如何保证大跨度大截面变曲率跨层桁架的施工精度是一项亟需解决的难题，本文从设计、加工、安装、焊接、卸载、监测各个环节来采取措施，保证施工精度的要求。
关键词：大跨度；大截面；变曲率；桁架；精度

Construction precision control measures and methods of large span and large section variable curvature span truss

Dong Tao Zhang Wenbin Zhang Yukuan Ma Xiaowei He Bin
(China Construction Eighth Engineering Bureau Steel Structure
Engineering Company, Shanghai 200120, China)

Abstract: With the rapid development of China's society and economy, personalized, unique and meaningful steel structure buildings are constantly emerging. Large span and large section variable curvature cross-layer truss steel structure is applied to building structure because of its novel shape and good structural strength stability. How to ensure the construction accuracy of large-span and large-section variable curvature cross-layer truss is an urgent problem to be solved, This article from the design, processing, installation, welding, unloading, monitoring each link to take measures to ensure the construction accuracy requirements.
Keywords: large span; large section; variable curvature; truss; accuracy

可交互式建筑资源智能引擎平台建设及应用

邓 洋 许 宁 黄文杰
（中国建筑第五工程局有限公司，湖南 长沙，410000）

摘 要：建筑行业作为国民经济的支柱产业之一，是国民经济的重要组成部分。近年来，行业经历了高速发展，知识密集增长，成果显著、范围宽广、类型繁多，同时也面临了知识产权保护、知识资源流失、知识价值难以发挥等问题。当前，在物联网、大数据、人工智能等信息技术广泛应用的时代背景下，知识的数字化转型成为企业发展的必经之路，本文依托企业科技成果管理项目实践，开发可交互式建筑资源智能引擎平台，可实现知识资源交互式数字化管理，促进科技资产增值，提高行业资源整合效益，助推行业高质量发展，为促进科研成果转化为生产力提供一种管理创新案例借鉴。
关键词：可交互式；建筑资源；智能引擎平台

Interactive building resource intelligent engine platform construction and application

Deng Yang Xu Ning Huang Wenjie
(China Construction Fifth Engineering Division Co., Ltd., Changsha, Hunan 410000, China)

Abstract: As one of the pillar industries of the national economy, the construction industry is an important material construction department of the national economy. In recent years, the industry has experienced rapid development, intensive growth of knowledge, remarkable achievements, wide range, various types, but also faced with intellectual property protection, the loss of knowledge resources, knowledge value is difficult to play and other problems. At present, in the Internet of Things, big data, artificial intelligence and other widely used under the background of the era of information technology, the digital transformation of knowledge into enterprise development path, relying on scientific and technological achievements management project practice, this paper develop a building interactive resources intelligent engine platform, which can realize interactive digital management of knowledge resources, promote the assets value of science and technology, Improve the industry resource integration benefits, boost the high-quality development of the industry, and provide a management innovation case for promoting the transformation of scientific research achievements into productivity.
Keywords: interactive; building resources; intelligent engine platform

基于知识细粒度提取及阅读技术的数字化成果平台建设研究

许　宁　邓　洋　黄文杰
（中国建筑第五工程局有限公司，湖南 长沙，410000）

摘　要：创新是引领发展的第一动力，企业作为科技创新的主体地位正在不断强化，众多企业依靠科技创新和技术进步加快传统劳动密集型产业的改造升级，实现发展方式根本性转变，其创新过程中各类科研成果成为提升企业实力和竞争力的重要组成。目前，大部分企业科研成果资源阅读和学习都还只是停留在对传统的PDF、WORD文档的版式阅读和利用，类似阅读器，不能对知识资源进行深度结构化加工及分类管理，研究一种基于知识细粒度提取及阅读技术在进行数字化成果平台建设中尤为重要。

关键词：知识细粒度提取及阅读技术；数字化成果平台

Research on construction of digital achievement platform based on fine-grained knowledge extraction and reading technology

Xu Ning　Deng Yang　Huang Wenjie
(China Construction Fifth Engineering Division Co., Ltd., Changsha, Hunan 410000, China)

Abstract: Innovation is the first power, leading the development of enterprises as the main body of technological innovation status is increasingly strengthened, and many enterprises rely on scientific and technological innovation and technological progress to speed up the upgrade of traditional labor-intensive industries, realize fundamental transformation development way, all kinds of scientific research achievements in the process of its innovation become an important enterprise strength and competitiveness. At present, most of the enterprise scientific research resources to read and learn is just to stay in the traditional PDF and WORD document format to read and use, similar to the reader, not on the depth of knowledge resources for structured processing and classification management, study a fine-grained extraction based on knowledge and reading technique in the construction of digital platform is particularly important.

Keywords: fine-grained knowledge extraction and reading technology; digital achievement platform

基于 DRM 技术的企业科技成果数字版权管理及保护

黄文杰　许　宁　邓　洋

（中国建筑第五工程局有限公司，湖南 长沙，410000）

摘　要：随着信息化技术的快速发展，DRM 数字版权管理技术在文本、图像、视频等数字化内容的管理及保护上得到了广泛的应用。本文以 DRM 技术为基础，根据建筑行业企业科研成果版权管理的需求，设计了企业科研成果文档数字版权管理及保护系统。系统使用特殊的 DRM 文件加密格式，可以配置相关权限模板及加密验证方式，能够有效地实现企业科研成果文档的数字版权保护。确保企业科研成果资源在合理、合法范围内传播利用，促进知识资源的业内共享。

关键词：DRM 技术；科研成果；数字版权管理；数字版权保护

Digital rights management and protection of enterprise scientific and technological achievements based on DRM technology

Huang Wenjie　Xu Ning　Deng Yang

(China Construction Fifth Engineering Division Co., Ltd., Changsha, Hunan 410000, China)

Abstract: With the rapid development of information technology, DRM technology has been widely used in the management and protection of digital contents such as text, image and video. Based on DRM technology, this paper designs a digital rights management and protection system for enterprise scientific research achievements documents according to the requirements of copyright management for construction enterprises. The system uses a special DRM file encryption format, and can configure the relevant permission template and encryption verification method, which can effectively realize the digital copyright protection of enterprise scientific research achievements documents. Ensure the dissemination and utilization of enterprises' scientific research achievements resources within a reasonable and legal scope, and promote the sharing of knowledge resources within the industry.

Keywords: DRM technology; scientific research achievements; Digital Rights Management; digital copyright protection

基于 BIM 的双向扭曲连续钢箱梁深化施工技术

孙晓伟　史　伟　尹建鲁　赵　阳　张少龙
（中建八局钢结构工程有限公司，上海，201304）

摘　要：二环北路及东西延伸段智慧快速路水上钢箱梁为空间双向扭曲结构，采用正交异性桥面结构。利用 CATIA 和 AUTOCAD 建立 BIM 模型，将模型进行纵向分段和横向分片。利用 IGS 三维数模，在 CATIA 和 AUTOCAD 之间时间模型数据的共享。并且利用 CATIA 的创成式曲面建模和创成式曲面展开功能，能够将空间曲面进行压线展开。能够确保钢箱梁几何精度、焊接质量，施工安全可靠，能够提高安装效率，确保钢箱梁和线型符合设计和规范要求。
关键词：水上钢箱梁；正交异性；CATIA；AUTOCAD；IGS；创成式曲面设计；压线展开

Deepening construction technology of bi-directional twisted continuous steel box girder based on BIM

Sun Xiaowei　Shi Wei　Yin Jianlu　Zhao Yang　Zhang Shaolong
(China Construction Eighth Engineering Bureau Co., Ltd., Shanghai 201304, China)

Abstract: The steel box girder on the water of the North Second Ring Road and the east-west Extension Section of the smart expressway is a spatial two-way twisted structure with orthotropic deck structure. Using CATIA and AUTOCAD to build BIM model, the model is divided into vertical section and horizontal section. The sharing of time model data between CATIA and AUTOCAD using IGS 3d digital model. And with the function of Catia's generative surface modeling and generative surface developing, the space surface can be expanded with lines. It can ensure the geometric precision, welding quality, construction safety and reliability, can improve the installation efficiency, and ensure that the steel box girder and line meet the design and specification requirements.
Keywords: steel box girder on water; orthotropic; CATIA; AUTOCAD; IGS; generative shape design; expanded with lines

基于混合架构的基坑监测信息管理系统设计与实现

张清明[1]　徐　帅[2]　杨　磊[1]
（1. 黄河水利委员会黄河水利科学研究院，河南 郑州，450003；
2. 黄河水资源保护科学研究院，河南 郑州，450003）

摘　要： 随着城市现代化进程的不断加快，基坑支护技术发展对基坑监测信息化管理提出了更高的要求。结合当前基坑监测现状，集合 C/S 架构和 B/S 架构的优势，采用 JavaEE 体系研发了基于混合架构的基坑监测信息管理系统，并对系统架构、功能模块、关系型数据库做了详细设计。系统架构包括基于 B/S 架构的公共管理平台、单位管理平台和基于 C/S 架构的采集客户端。功能模块包括机构信息、工程信息、监测数据、报告管理和系统管理等。平台和客户端共用 MYSQL5. 0 关系型数据库，可实现平台、客户端数据的向上同步与向下同步，实现从监测数据采集、分析处理、上传、预警到监测报告编制、导出的自动化，以及图形与属性、图形与监测信息的联动，有效提高了基坑监测信息化管理的水平和工作效率。

关键词： 混合架构；基坑；监测；管理系统

Design and implementation of foundation pit monitoring information management system based on hybrid architecture

Zhang Qingming[1]　*Xu Shuai*[2]　*Yang Lei*[1]
（1. Yellow River Institute of Hydraulic Research，YRCC，Zhengzhou，Henan 450003，China；
2. Yellow River Water Resources Protection Research Institute，
Zhengzhou，Henan 450003，China）

Abstract： With the accelerating process of urban modernization，the development of foundation pit support technology has put forward higher requirements for the informatization management of foundation pit monitoring. Combined with the current status of foundation pit monitoring，the advantages of C/S architecture and B/S architecture are integrated，and the foundation pit monitoring information management system based on hybrid architecture is developed by using JavaEE system. The system architecture，function modules and relational database are designed in detail. The system architecture includes the public management platform based on B/S architecture，the unit management platform and the acquisition client based on C/S architecture. The function modules include institutional informa-

tion, engineering information, monitoring data, report management and system management. The platform and client share the MYSQL5. 0 relational database, which can realize the upward and downward synchronization of the data of the platform and client, and realize the automation of monitoring data acquisition, analysis and processing, upload, early warning, preparation and export of monitoring reports.
Keywords: hybrid architecture; foundation pit; monitoring; management system

超高层结构竖向变形与伸臂桁架安装时序研究

田喜胜[1]　简宏儒[1]　冯　吉[1]　王　军[1]　廖　继[1]　范海峰[2]　谢卓霖[2]
（1. 中建三局第三建设工程有限责任公司（西南），重庆，400000；2. 重庆大学，重庆，400044）

摘　要：本文结合材料的收缩徐变理论，对带有伸臂桁架的超高层进行施工模拟分析，得到了施工过程中材料收缩徐变对结构的影响，探讨了竖向变形的影响因素和伸臂桁架的安装时序。研究结果对超高层结构竖向变形变化规律的掌握具有参考价值，所研究的伸臂桁架安装时序对后续带有伸臂桁架的巨柱核心筒超高层结构施工具有指导意义。
关键词：超高层结构；收缩徐变；竖向变形；伸臂桁架

Study on vertical deformation of super high rise structure and installation sequence of outrigger truss

Tian Xisheng[1]　*Jian Hongru*[1]　*Feng Ji*[1]　*Wang Jun*[1]　*Liao Ji*
Fan Haifeng[2]　*Xie Zhuolin*[2]
(1. The Third Construction Engineering Co., Ltd. (southwest), Chongqing 400000, China;
2. Chongqing University, Chongqing 400044, China)

Abstract: In this paper, combined with the theory of material shrinkage and creep, the construction simulation analysis of super high-rise buildings with outrigger truss is carried out, and the influence of material shrinkage and creep on the structure during the construction process is obtained, and the influencing factors of vertical deformation and the installation of outrigger truss are discussed. Timing. The research results are of reference value for the mastery of the law of vertical deformation of super high-rise structures. The researched outrigger truss installation sequence has guiding significance for the subsequent super high-rise construction of giant column core tube structure.
Keywords: super high-rise structure; shrinkage and creep; vertical deformation; outrigger truss

装配式智能化泵组模块应用技术

张　超　吴　刚　李湖辉　王礼杰　张　琛
（中建五局第三建设有限公司，湖南 长沙，410004）

摘　要：以合肥轨道4号线项目制冷机房为例，在装配式机电技术的基础上采用智能化的预制方式，将标准化泵组模块进行强电、群控系统的功能拓展，解决阀组及管道构件设计的多样不统一，模块构件的系统防水、防潮、防噪、恒温、设备运行情况监控、远程群控等工程集成等难题，实现了装配式智能化泵组模块的流水化生产、智能化运维。
关键词：装配式机电；智能化泵组模块；在线监测；智能化运维

Application technology of assembled intelligent pump group module

Zhang Chao　Wu Gang　Li Huhui　Wang Lijie　Zhang Chen
(The Third Company of China Fifth Engineering Bureau Ltd, Changsha, Hunan 410004, China)

Abstract: Taking the refrigerating room of Hefei railway line 4 project as an example, on the basis of assembled mechanical and electrical technology, standardized pump set of modules was improved with function expansion of strong electricity and group control system by the way of intelligent prefabrication, so as to solve the diverse design of valve set and pipeline components, and the integrated engineering problem of water and moisture resistance, noise reduction, constant temperature, equipment operation monitoring, long-distance control of module component system, and to realize the flow production and intelligent operation and maintenance of the assembled intelligent pump group module.
Keywords: assembled mechanical and electrical; intelligent pump module; on-line monitoring; intelligent operation and maintenance

BIM 技术在香港机场天际走廊大跨度钢结构桥梁工程中的应用

高　翔　刘裕禄
（中国建筑工程（香港）有限公司，中国 香港，999077）

摘　要： 天际走廊项目位于香港国际机场运营中的禁区停机坪内，工期紧，安全风险高，施工难度大。项目团队采用整体建模计算方法及施工全过程时变分析仿真技术，通过建立BIM工作组织以指导工序安排与进度管理、钢结构设计生产及管理、钢结构桥梁拼装、主桥大节段钢结构运输、主桥整体运输与顶升、施工过程力学分析，解决施工中的技术难题和管理难题，确保主桥钢构件高精度生产、高精度拼装，推进主桥运输、顶升、安装按计划顺利完成，为BIM技术在工程实践中的应用探索提供又一先进典型案例。本文结合该工程中的具体实践经验，详细介绍BIM技术在大跨度钢结构桥梁施工中的应用。
关键词： BIM；钢结构；大跨度桥梁；施工模拟

Application of BIM technology in construction of long-span steel structure bridge of Hong Kong international airport sky bridge

Gao Xiang　Liu Yulu
(China State Construction Engineering (Hong Kong) Limited, Hong Kong 999077, China)

Abstract: The Sky Bridge project is located in the restricted apron of the Hong Kong International Airport in operation, of which the construction period is tight, the safety risk is high, and the construction is difficult. The project team adopts integral modeling and calculation method and time-varying analysis and simulation technology for the whole construction process, then establishes a BIM work organization to guide process arrangement and schedule management, steel structure design, production and management, steel structure bridge assembly, and main bridge large section steel structure transportation, the overall transportation and lifting of the main bridge, and the mechanical analysis of the construction process, to solve the technical and management problems in the construction, so that ensure the high-precision production and high-precision assembly of the steel components of the main bridge, and promote the smooth completion of the main bridge transportation, lifting and installation as planned. This provides another advanced typical case for the exploration of the application of BIM technology in engineering practice. This article combines specific practical experience in the project to introduce in detail the application of BIM technology in the construction of long-span steel structure bridges.
Keywords: BIM; steel structure; long-span bridge; construction simulation

BIM 技术在低能耗建筑设计中的协同作用
——以建筑方案设计优化为例

凌　晨　何仁儒

（中海地产集团南昌公司）

摘　要： 走绿色化之路是我国建筑行业发展的主旋律，而设计阶段是使绿色建筑得以落现的第一步也是最重要的一步。本文利用响应面设计方法做试验设计与分析，结合建筑信息模型及建筑性能模拟软件的耗能指标分析结果，以此得到建筑耗能设计影响因子的显著性程度。研究发现墙传热系数与窗传热系数对能耗指标影响显著；窗墙比和屋面传热系数的 P 值大于 0.05，说明窗墙比和屋面传热系数对能耗指标影响不显著。此外，窗墙比和屋面传热系数的交叉作用 P 值小于 0.05，说明两者交叉作用对能耗指标影响显著。通过模型的计算优化，得到能耗指标最少的热环境下的模拟参数设定，即窗墙比 0.50，墙传热系数 0.67，窗传热系数 5.50，屋面传热系数 0.73。本文可为以节能为方向的建筑耗能设计参数选取提供借鉴。

关键词： BIM；低能耗建筑；协同作用；优化；绿色建筑

The synergistic effect of BIM technology in the design of low energy building
—— Taking the optimization of building scheme design as an example

Ling Chen　He Renru

Abstract: Going green is the inevitable trend of the development of China′s construction industry, and the design stage is the first and most important step to realize the green building. In this paper, response surface design method and building information modeling (BIM) and building performance simulation software are used to analyze the significance level of architectural design influencing factors, which is convenient to guide the parameter selection design with energy saving as the goal. It is found that the heat transfer coefficient of wall and window have significant influence on the energy consumption index. The P values of window-wall ratio and roof heat transfer coefficient are greater than 0.05, indicating that the window-wall ratio and roof heat transfer coefficient have no significant effect on energy consumption index. In addition, the P value of the cross effect between the window-wall ratio and the roof heat transfer coefficient is less than 0.05, indicating that the cross effect of the two has a significant impact on the energy consumption index. Through

the calculation and optimization of the model, the simulation parameter Settings under the thermal environment with the least energy consumption index were obtained: window-wall ratio 0. 50, wall heat transfer system, 0. 67, window heat transfer system, 5. 50, roof heat transfer system, 0. 73.
Keywords: BIM; low-energy buildings; synergy effect; optimization; the green building

关于BIM技术在大跨径梁拱组合刚构桥项目中的前期实施策划研究

王　蓬　李亚勇
（中建隧道建设有限公司，重庆）

摘　要：根据国家“十四五”规划的部署，建筑业信息化势在必行，住房和城乡建设部等十三部门发布的《关于推动智能建造与建筑工业化协同发展的指导意见》确定了中国将走新型建筑工业化之路，并确立了“中国建造”的战略地位。随着BIM技术在进入施工实践领域深水区，BIM应用也向施工现场领域延伸与发展，而BIM的实施策划无疑是BIM技术在施工现场落地的重要部分。
关键词：BIM技术；施工现场；BIM实施应用策划

Research on the early implementation planning of BIM technology in long span beam arch composite rigid frame bridge project

Wang Peng　Li Yayong
(China construction Tunnel Co., Ltd., Chongqing, China)

Abstract: According to the deployment of the national 14th five year plan, the informatization of the construction industry is imperative. The guiding opinions on promoting the coordinated development of intelligent construction and building industrialization issued by the Ministry of housing and urban rural development and other 13 departments confirmed that China will take the road of new building industrialization, and established the strategic position of " built in China"[1]. With BIM Technology entering the deep water area of construction practice, BIM application also extends and develops to the field of construction site, and BIM implementation planning is undoubtedly an important part of BIM Technology landing in the construction site.
Keywords: BIM technology; construction site; BIM implementation and application planning

BIM 技术在超大型国际展馆钢结构施工中的应用

龙　攀　田　华　戴　秘　杨飞凤　龙婉东　李　智
（中国建筑第五工程局有限公司）

摘　要：近几年国家公共建筑大力发展，各省市展厅场馆建的也越来越多越来越大，成为城市发展的重要组成部分，但大型场馆施工时由于本身施工场地广、结构复杂、施工工期紧、构件数量多、专业交叉频繁等特点，在施工中往往会遇到质量把控难度大、安全隐患多等顾此失彼的情况。威海国际经茂交流中心项目结合施工实际情况，大力运用 BIM 技术为多结构施工排忧解难，积累了不少运用经验，现笔者将其整理成文与广大读者分享。
关键词：超大型展馆；BIM 技术；施工管理；多结构施工

浅谈华南理工大学广州国际校区二期（第一批次）项目BIM技术在项目实施过程中的应用

徐为 马川 马凯 田华 戴秘
（中国建筑第五工程局有限公司，广东 广州，511400）

摘　要：近几年来，随着国家政策的大力推行，建筑信息模型（Building Information Modeling，BIM）技术逐渐在施工单位、设计单位普及，但BIM技术在设计、设计管理、施工管理等过程的应用深化及应用点存在大量"PPT·BIM"，并未有效地在过程中提供明确的可量化的指标内容。本着BIM技术在EPC模式下的管理应用，发掘深思在工程项目过程中对人力、物力、财力的集约化、可视化、信息化的能力，探索BIM技术与管理的融合。本文主要针对BIM技术在华南理工大学广州国际校区二期（第一批次）设计施工总承包项目的具体应用与管理探讨和研究。

关键词：BIM技术；施工管理；传统施工；智慧建造

专题五　智　慧　城　市

基于大数据之下智慧城市建设模式思考

李青宇
（中建五局第三建设有限公司，湖南 长沙，410007）

摘　要： 随着移动互联网、物联网在人们实际生活中的广泛应用，在这个大数据快速发展的时代，智慧新型城市的管理建设仍然值得人们投入更深入、多层次的思考，如何有效应对这些大规模、复杂的智慧城市数据库也面临着巨大的新挑战。笔者从一个综合性的视角深入思考了在大数据技术驱动下我国智慧城市发展建设过程中的若干重要问题，并对这些问题提出具有针对性的解决建议。

关键词： 大数据；智慧城市；思考；问题

Thought on building pattern of smart city based on big data

Li Qingyu
(3rd Construction Co., Ltd. of China Construction 5th Engineering Bureau,
Changsha, Hunan 410007, China)

Abstract: Along with the wide applications of mobile internet and the internet of things in human real life, the management and building of the new type of smart city is worth deep and multilevel thinking. How to efficiently deal with such large-scale and complicated database of smart city is a great challenge to us. The author has considered a series of important issues of the Chinese smart city's building and development which are driven by the big data technology. Furthermore, the author also proposed some corresponding suggestions to the above issues.

Keywords: big data; smart city; thinking; problems

新型冠状病毒肺炎疫情影响下的天津地铁客流分析及运营组织研究

王多龙[1] 刘庆磊[1] 尹浩东[2] 王鹤天[3]

（1. 天津一号线轨道交通运营有限公司，天津，300022；

2. 北京交通大学 交通运输学院，北京，100044；

3. 北京京投亿雅捷交通科技有限公司，北京，100044）

摘　要： 新冠肺炎疫情的暴发，使城市轨道交通运营管理工作面临着严峻的挑战，不仅要做好员工和乘客的疫情防控工作，还要做好日常轨道交通运输组织工作。本文从路网、线路和站点三个层级，纵向对比往年同期，分析新冠肺炎疫情下的天津地铁客流特征。力求对防疫复工期间地铁运行突出特点进行精细刻画。在客流分析基础上，总结边运营边防疫经验，为处置公共卫生事件提供参考。

关键词： 新冠肺炎疫情；城市轨道交通运营管理；客流特征；天津地铁；公共卫生事件

Analysis of passenger flow and operation organization in Tianjin Metro under the influence of COVID-19

Wang Duolong[1] *Liu Qinglei*[1] *Yin Haodong*[2] *Wang Hetian*[3]

（1. Tianjin Metro Line 1 Rail Transit Operation Limited Company，Tianjin 300022，China；

2. Beijing Jiaotong University School of Traffic and Transportation，Beijing 100044，China；3. Beijing BII-ERG Transportation Technology Co.，Ltd.，Beijing 100044，China）

Abstract：After the outbreak of COVID-19 epidemic，the operation and management of urban rail transit has facing severe challenges. It is necessary not only to prevent and control the epidemic for employees and passengers，but also to organize daily rail transit transportation. This article analyzed the characteristics of Tianjin Metro passenger flow under the COVID-19 epidemic from three levels of road network，line and station，and analyzed the characteristics of Tianjin Metro passenger flow under the COVID-19 epidemic. The purpose is striving to finely describe the outstanding characteristics of metro operation during the resumption of epidemic prevention. On the basis of passenger flow analysis，summed up the experience of as operations as epidemic prevention to provide reference for handling public health incidents.

Keywords：COVID-19 epidemic；urban rail transit operation management；passenger flow characteristics；Tianjin metro；public health event

北京大兴国际机场轨道线航空旅客行李托运系统实践

马立秋[1]　金　奕[2]　胡家鹏[3]　赵永康[2]　梅　棋[3]

（1. 北京市轨道交通建设管理有限公司，北京，100068；
2. 北京城市铁建轨道交通投资发展有限公司，北京，100068；
3. 北京市轨道交通设计研究院有限公司，北京，100068）

摘　要：为作为机场在城市的延伸——城市航站楼，在提高公共交通出行比例，减少机场航站楼的压力、提高城市品质等方面能发挥积极作用。城市航站楼的核心功能之一是为旅客提供行李托运，本文在广泛调研的基础上，详述了大兴机场线行李托运系统的设计及对比分析，为以后我国城市航站楼行李托运系统的建设提供参考。

关键词：行李托运系统；大兴机场轨道线；行李小车；渡板

Practice of luggage check-in system on Beijing Daxing International Airport Rail Line

Ma Liqiu[1]　*Jin Yi*[2]　*Hu Jiapeng*[3]　*Zhao Yongkang*[2]　*Mei Qi*[3]

（1. Beijing Metro Construction Administration Co., Ltd., Beijing 100068, China;
2. Beijing CSTJ Metro Investment Development Co., Ltd., Beijing 100068, China;
3. Beijing Metro Design and Research Institute Co., Ltd., Beijing 100068, China）

Abstract: As an extension of the airport in the city, the city terminal play a positive role in improving the proportion of public transport, reducing the pressure of the airport terminal, and improving the quality of the city. One of the core functions of the city terminal is to provide luggage check-in. On the basis of extensive investigation, this paper describes the design and comparative analysis of the luggage check-in system of Daxing airport line, which provides a reference for the future construction of the luggage check-in system of the city terminal in China.

Keywords: luggage check-in system; daxing airport line; luggage trolley; cab apron

超高层悬挑结构日照变形研究

武传仁　王得明　李善文　卞　吉　衡成禹
（中国建筑第八工程局有限公司，上海，200000）

摘　要：随着超高层建筑应用逐渐广泛，为实现造型或建筑功能需求，悬挑结构应运而生。因太阳辐射不均匀和日照温差不同，对超高层全钢结构建筑施工有较大影响。以扬子江国际会议中心建设项目酒店区塔楼为背景，采用全站仪和应变设备对塔楼进行监控测量。然后对采集数据进行分析，研究日照对超高层悬挑结构变形影响。研究表明，超高层建筑钢框架悬挑结构施工，随施工时间推移，日照温度变化较为明显，需要根据不同施工期内，大环境中温度变化天预留结构变形。同时避免在高温环境下进行定位。

关键词：悬挑结构；超高层；日照变形；全站仪

Study on sunshine deformation of super high rise cantilever structure

Wu Chuanren　Wang Deming　Li Shanwen　Bian Ji　Heng Chengyu
(China Construction Eghth Engineering Division Co., Ltd., Shanghai 200000, China)

Abstract: With the increasingly wide application of super high-rise buildings, cantilever structure emerges as the times require in order to realize the modeling or building function requirements. Because of the uneven solar radiation and different sunshine temperature difference, it has a great influence on the construction of super high-rise steel structure. Taking the hotel tower of Yangtze River International Conference Center construction project as the background, total station and strain equipment are used to monitor and measure the tower. Then the collected data are analyzed to study the influence of sunshine on the deformation of super high-rise cantilever structure. The research shows that, in the construction of steel frame cantilever structure of super high-rise building, the sunshine temperature changes obviously with the construction time, so it is necessary to reserve the structural deformation according to the temperature change days in different construction periods. At the same time, avoid positioning in high temperature environment.

Keywords: cantilever structure; super high rise; sunshine deformation; total station

跨孔地震层析成像技术及全天候自动监测系统在澳门老旧建筑周边开展工程施工中的应用研究

冯少磊　霍辰君　阮君豪　劳永权　贲亦骁
（中国建筑工程（澳门）有限公司，中国 澳门，999078）

摘　要：在澳门老旧城区开展的都市更新工程，特别是在地基、基础和基坑开挖阶段，对于工程周边的老旧建筑需要采取技术措施保障其安全性。通过研究将跨孔地震层析成像及全天候建筑物自动监测系统，应用于澳门老旧建筑周边的工程项目中，保障项目周边建筑物的安全。项目采用跨孔地震层析成像技术，探测新建项目周边老旧房屋的地基土质状况，根据所得地质报告优化施工方案；项目采用全天候建筑物自动监测系统监测周边老旧建筑的状况，保障施工时周边老旧房屋的结构安全。研究表明，跨孔地震层析成像技术的应用可取得直观可靠的数据并模拟出三维模型，以便在没有条件进行钻探的情况下了解地质情况；建筑物自动监测系统是将物联网技术应用于建筑物监测的新型监测系统，可通过自动化的数据测量、网络传输和云端数据处理系统形成的物联网，随时监测到建筑物的变化情况，在节省劳动力的同时能通过日常通信设备即时反馈周边建筑物的安全状况。有关技术在未来的澳门都市更新工程中将具有很高的推广价值和应用效益。
关键词：跨孔地震层析成像；建筑物自动监测系统；物联网；澳门都市更新工程

The study of cross-hole seismic tomography and all-weather automatic building monitoring system application in the construction project surrounded by old buildings in Macau

Feng Shaolei　Huo Chenjun　Un Kuanhou　Lo Wingkuen　Ben Yixiao
(China State Construction Engineering (Macau) Company Ltd., Macau 999078, China)

Abstract: To start the urban renewal project surrounded by old buildings in Macau, the safety of surrounded old buildings should be ensured by technical measures especially in the phase of foundation-basement-excavation stage. Cross-hole seismic tomography and all-weather automatic building monitoring system are applied on the project surrounded by old buildings in Macau. Cross-hole seismic tomography was used to enhance the construction method according to the conducted report of the geologic condition under the surrounding old buildings. All-weather automatic building monitoring system was used in the project to ensure the safety of the surrounding old buildings by monitoring the condition of surround-

ing buildings. The study shows that the application of cross-hole seismic tomography can obtain sufficient and visualized information and generate the 3-D model more conveniently while the location is unable to investigate the underground condition by drilling method. Automatic building monitoring system, a new monitoring system of application of Internet of Things, is able to monitor the changes of the buildings at all times via the IoT composed of automatic data measuring, network transmission and cloud data processing; besides, save in labor and ensure the immediate safety condition of surrounding buildings via common communication equipment. The related technologies will have high promotion value and application benefits in the future urban renewal projects in Macau.
Keywords: cross-hole seismic tomography; automatic building monitoring system; Internet of Things; Macau urban renewal project

快速轨道交通工程防淹防护密闭门技术研究

王阳明　胡圣伟　熊　科　贾彦明　陶　涛
（广州地铁设计研究院股份有限公司，广东 广州，510010）

摘　要：在全球水旱灾害极端事件频发和我国地下空间规模快速发展的背景下，为提高地铁隧道平时及战时的防灾减灾能力，针对地下隧道防淹设施的使用环境、功能及性能要求，对防淹防护密闭门系统及其关键结构开展了相关技术研究。为保证防淹防护密闭门的防水密闭要求，提升其运行稳定性和可靠性，降低其安装调试及维护难度，研制了适于地铁隧道环境的防淹防护密闭门，经工程实践验证，功能完善，性能优良，运行稳定可靠，安装及维护方便，极具推广和应用价值。本文就所研制的防淹防护密闭门系统组成、功能、实现原理和结构特点等进行相关介绍。

关键词：地铁；电动升降式；液压立转式；防淹防护密闭门

Study on flood gate and airtight blast door in rapid rail transit project

WangYangming　Hu Shengwei　Xiong Ke　Jia Yanming　Tao Tao
(Guangzhou Metro Design & Research Institute Co., Ltd.,
Guangzhou, Guangdong 510010, China)

Abstract: Under the background of frequent extreme events of flood and droughts in the world and the rapid development of underground space in China, in order to improve the disaster prevention and mitigation ability of subway tunnels in peacetime and wartime, according to the use environment, function and performance requirements of underground tunnel anti-flood facilities, the relevant technical research on the flood gate and airtight blast door system and its key structures is carried out. In order to ensure the waterproof and sealing requirements of the flood protection airtight blast door, improve its operation stability and reliability, and reduce the difficulty of installation, debugging and maintenance, the flood protection airtight door suitable for subway tunnel environment was developed. It has been verified by engineering practice that it has perfect functions, excellent performance, stable and reliable operation, convenient installation and maintenance, and is of great promotion and application value. In this paper, the composition, function, realization principle and structural characteristics of the developed flood gate and airtight blast door system are introduced.

Keywords: metro; motor-driven raising; hydraulic vertical rotary; flood gate and airtight blast door

基于噪声辨识技术的装配式智慧城市环境监测系统研究

邢　晨[1,2]　余　磊[1]　王　波[2]　王　玺[2]

（1. 哈尔滨工业大学（深圳），广东 深圳，440300；

2. 中建科技集团有限公司，广东 深圳，440300）

摘　要： 目前传统噪声监测技术主要以环境声级作为监测指标，然而随着健康城市理念的发展，这一指标已不能反映城市建设过程中声音环境的全貌。在解决城市建设过程中的噪声问题上，除了控制环境声级外，还需针对性地处理不利声源，提升有利声源。本研究从使用者感知的角度出发，通过对环境声源特征分析，进行了不同环境噪声的“噪度”分类及识别工作，分别构建了基于环境声“烦躁度”的噪声声源辨识技术，以及基于自然声比率识别的环境噪声源辨识技术。最终借助噪声监测设备，构建了可实时进行噪声监测、识别和可视操作的一体化的装配式监测系统，实现城市中尺度区域范围的智慧环境监测。

关键词： 装配式基站；噪声监测；噪声源识别；智慧城市监测系统

Research on prefabricated smart city environmental monitoring system based on noise identification technology

Xing Chen[1,2]　*Yu Lei*[1]　*Wang Bo*[2]　*Wang Xi*[2]

（1. Harbin Institute of Technology Shenzhen，Shenzhen，Guangdong 440300，China；

2. China Construction Technology Group corporation，Shenzhen，Guangdong 440300，China）

Abstract： At present，traditional noise monitoring technology mainly uses environmental sound level as main indicator. However，with the development of the concept of healthy cities，this method can no longer reflect the full picture of the sound environment. In order to solve the noise problem in the process of urban construction，attenuating absolute sound level for urban noise control is not always efficient in which sound meanings has to be included. Based on user perception，this research carried out the classification of different noises through the analysis of the characteristics of the environmental sound source，and constructed the noise recognition technology based on the sound annoyance evaluation. And constructed a noise source identification technology based on the identification of natural sound proportions. Finally，an integrated prefabricated monitoring system that can perform real-time noise monitoring and visual operation is built by use noise monitoring equipment. It can realize smart environmental monitoring in the middle-scale area of the city.

Keywords： prefabricated base station；noise monitoring；noise source identification；smart city monitoring system

基于适风设计的高层建筑智能覆面构件

柯延宇　沈国辉　杨肖悦　谢霁明
（浙江大学 建筑工程学院，浙江 杭州，310058）

摘　要：本文以竖向肋板为例，阐述了覆面构件对结构风致响应的影响。对四种建筑覆面构件进行了风洞试验，计算了加速度、倾覆力矩和基底剪力的风致响应，可知相对宽度为4%的竖向肋板所产生的气动干扰可能会破坏涡的规则脱落，并导致建筑物的横风振动明显小于光滑工况和相对宽度为2%的工况。此外，布置在角区的半分布肋条与全分布肋条的风致响应无显著差异，这说明角区肋条对减小不利气动力产生的风响应起着关键作用。通过覆面构件的变化减小了风荷载阐释了适风设计的概念。建筑覆面在日常风速下可以维持光滑状态，或者为了一定的视觉效果和美学功能，可以伸出2%相对宽度的肋条。可以在建筑物表面配备分布式传感器，通过风速计和局部气象数据来获取当前建筑承受风速，并预测未来一段范围内的风速变化趋势，进而更好地调整覆面构件的布置情况。为了更好降低成本，在建造过程当中，可以只对角部区域预装竖向肋板，当强风来临时候让角部区域的肋板伸出相对宽度4%即可达到降低风致响应的效果。

关键词：适风设计；高层建筑；覆面构件；竖向肋板；风致响应

Intelligent cladding components of high rise building based on wind-adaptable design

Ke Yanyu　Shen Guohui　Yang Xiaoyue　Xie Jiming
（College of Civil Engineering and Architecture，Zhejiang University，
Hangzhou，Zhejiang 3150058，China）

Abstract：In this paper，the vertical ribs was used to discuss the influence of cladding components on the wind-induced response of the structure. Four types of model configurations were carried out in the wind tunnel，and the wind responses of acceleration，overturning moment and base shear force were calculated. It can be concluded that the aerodynamic interference due to the vertical ribs with 4% relative width may have a beneficial effect of disrupting the regular shedding of vortices and causing the across-wind oscillation of buildings appreciably smaller than that of the smooth one and the 2% relative width. In addition，there were no significant distinctions between the arrangement of half-distributed ribs in corner region and full-distributed ribs. It indicated that the ribs in corner region played a vital role in reducing the wind responses produced by unfavorable aerodynamic corner. The concept of wind-adaptable design is introduced since the wind loads are mitigated by the

cladding components. The building surface can keep the smooth state under the daily wind speed, or extend 2% of the relative width of ribs for a certain visual effect and aesthetic function. Distributed sensors can be equipped on the surface of the building to obtain the current wind speed of the building through the anemometer and local meteorological data, and predict the wind speed trend in the future, so as to better adjust the layout of the components. In order to reduce the cost, the vertical ribs can be installed on the corner area in the construction process. When the strong wind comes, the rib plate in the corner area can be extended by 4% of the relative width to achieve the effect of reducing the wind-induced response.
Keywords: wind-adaptable design; high-rise buildings; cladding components; vertical ribs; wind-induced responses

重庆山区高速公路行车安全相关因素的探讨研究

王　军[1,2]　孙伟亮[1,2]　杨　勇[1,2]

（1. 安徽省交通规划设计研究总院股份有限公司，安徽 合肥，230088；
2. 公路交通节能环保技术交通运输行业研发中心，安徽 合肥，230088）

摘　要：在重庆山区高速公路的建设和运营中，一旦人、车、路和环境因素发生改变，车辆行驶安全可能会受到威胁。以重庆山区高速公路交通 607 组行车数据为例，建立广义有序逻辑回归（gologit）模型，探索影响车辆行驶安全的相关因素及其影响作用。结果表明，12 个解释变量，包括驾驶员年龄、性别、酒驾、超速、疲劳驾驶、车辆类型、车辆有故障、超载、道路几何线形、路面状况、气候条件和节假日，对行车安全的影响显著。而且，解释变量酒驾违背了平行线假设，研究成果对改善重庆山区高速公路车辆行驶安全提供了重要的技术支撑。

关键词：高速公路；行车安全；gologit 模型；平行线假设

Exploring the factors affecting traffic safety in mountain highway of Chongqing

Wang Jun[1,2]　*Sun Weiliang*[1,2]　*Yang Yong*[1,2]

(1. Anhui Transportation Consulting & Design Institute Company, Hefei, Anhui 230088, China; 2. Road Traffic Energy Conservation and Environmental Protection Technology Transportation Industry R&D Center, Hefei, Anhui 230088, China)

Abstract: With the construction and operation of mountain highway in Chongqing, once a person, vehicle, road and environmental factors changes, the traffic safety might be threatened. Collecting the 607 Chongqing 607 driving datas in Mountain Highway Traffic of chongqing, we established generalized ordered logistic regression (gologit) model to explore the factors affecting the safety of the vehicle and its influence. The results showed that 12 explanatory variables, including the driver's age, sex, drunk driving, speeding, fatigue driving, vehicle type, vehicle failure, overloading, road geometry alignment, road conditions, weather conditions and holidays, have the impact on traffic safety significantly. And the explanatory variables, drunk driving violated the parallel lines assumption, the research can improve traffic safety in mountainous highway of chongqing and provides important technical support.

Keywords: highway; traffic safety; gologit model; parallel lines assumption

铁路信号数据智能化定测控制系统的设计研究

张 望[1] 马 浩[1] 郑 军[2] 袁国堂[2]
(1. 中铁建电气化局集团第三工程有限公司，河北 高碑店，074000；
2. 浙江大学台州研究院，浙江 台州，318000)

摘　要：近几年，中国铁路事业发展迅速，铁路建设智能化成为热门课题。本文介绍了一种铁路信号工程数据定测及检测装置，利用先进的 RTK 定位系统、车轮里程计结合三维激光扫描建模定位，在指定位置对轨道电路电压、电流、载频和补偿电容进行自动测量，并在要求的设备安装位置区域进行喷码标记，同时进行轨旁信号设备建筑的限界测量并自动保存数据。该装置可以有效提高整体的测量精度、提升作业效率，保证施工安全和施工质量。

关键词：RTK 定位系统；三维激光扫描；轨道电参数；喷码标记；限界测量

Design and research of intelligent railway signal data measurement control system

Zhang Wang[1] *Ma Hao*[1] *Zheng Jun*[2] *Yuan Guotang*[2]
(1. The 3rd Engineering Co. Ltd. of China Railway Construction Electrification Bureau Group, Gaobeidian, Heibei 074000, China; 2. Research Institute of Zhejiang University-Taizhou, Taizhou, Zhejiang 318000, China)

Abstract: China's railway industry has developed rapidly in recent years, and intelligent railway construction has become a hot topic. This paper introduces a railway signal engineering data measurement and detection device, which using advanced RTK positioning system and wheel odometer combining 3d laser scanning and modeling positioning system to realize three main measurement functions: automatically measure the voltage, current, carrier frequency and compensation capacitance of the railway circuit at the specified position; mark the required equipment installation position area by spraying QR Code; measure the boundary of the railside signal equipment building and automatically saving the data. The device can effectively improve the overall measurement accuracy and operation efficiency, ensure construction safety and construction quality.

Keywords: RTK positioning system; 3d laser scanning; railway electrical parameters; code spraying; boundary measurement

专题六　现　代　桥　隧

基于时间离散模型拉索减振系统的主动时滞补偿与半主动控制对比研究

方　聪[1]　周　帅[1]　雷　军[1]　何昌杰[1]　李水生[1]　谭芝文[2]
（1. 中国建筑第五工程局有限公司，湖南 长沙，410011；
2. 中建隧道建设有限公司，重庆，401320）

摘　要：斜拉桥拉索减振是行业热点问题，通过对拉索施加轴向控制力改变其刚度、阻尼是一种有效的方法，系统时滞的存在会对减振效果产生影响。本文建立磁致伸缩作动器动力学模型和时间离散模型的拉索-磁致伸缩作动器面内控制系统方程，提出基于移相法的拉索减振时滞补偿方法，借助数值仿真模拟说明时滞对系统主动与半主动控制方式减振效果的影响，同时采用移相法对主动控制进行时滞补偿。研究表明，系统存在时滞的情况，主动控制方式下减振效果影响明显，而半主动控制方式却没有影响，通过移相法对拉索系统进行时滞补偿可以产生良好的减振效果，且主动方式要优于半主动方式。

关键词：拉索；轴向控制力；时滞；时间离散模型；减振

Comparative study on active delay compensation and semi-active control for cable vibration absorber system based on time discrete model

Fang Cong[1]　*Zhou Shuai*[1]　*Lei Jun*[1]　*He Changjie*[1]　*Li Shuisheng*[1]　*Tan Zhiwen*[2]
(1. China Construction Fifth Engineering Division Co., Ltd., Changsha, Hunan 410004 China; 2. China Construction Tunnel Co. Ltd., Chongqing 401320, China)

Abstract: Vibration reduction of cable-stayed Bridges is a hot issue in the industry. It is an effective method to change the stiffness and damping of cables by applying axial control force. The control effect will be affected by the time delay of the system. In this paper, the equations of cable-magnetostrictive actuator in-plane control system based on the dynamic model and time discrete model of the magnetostrictive actuator are established, and a time-delay compensation method based on phase-shifting method for cable vibration reduction is proposed. By means of numerical simulation, the influence of time delay on the vibration reduction effect of the active and semi-active control modes of the system is illustrated. At the same time, the phase shift method is used to compensate the time delay of active control. The results show that the active control mode has a significant effect on the vibration reduction effect when the system has time delay, while the semi-active control mode has no

difference. The time-delay compensation of cable system by phase-shifting method can produce good vibration reduction effect, and the active mode is better than the semi-active mode.
Keywords: the stay cable; axial control force; time delay; time discrete model; vibration reduction

Ⅴ级围岩隧道双侧壁导坑法开挖掌子面破坏特征研究

聂奥祥
（北京市市政工程设计研究总院有限公司，北京，100082）

摘　要： 掌子面失稳破坏是软弱围岩大断面隧道施工面临的主要工程难题之一，以依托工程为研究对象，采用三维有限元法就双侧壁导坑开挖时的掌子面破坏特征进行了研究，研究表明：1）在软弱围岩大断面隧道开挖过程中，仅仅通过超前支护很难达到保证掌子面稳定的目的，必须通过掌子面加固措施来提高掌子面的围岩物理力学参数，以增加掌子面稳定性。2）采用双侧壁导坑法施工时，各个导洞之间开挖引起的相互影响作用是不可忽视的，且距离越近，相互影响作用越明显。3）双侧壁导坑法施工时，中导洞上台阶掌子面为最容易失稳部位，施工时应引起特别关注。4）为防止施工过程中掌子面破坏不适当的扩大以及不同破坏区域之间的贯通导致隧道发生整体掌子面失稳现象，各导洞的开挖进尺应控制在 1.0m 以内，且不宜进行多个导洞同时开挖。

关键词： Ⅴ级围岩；大断面隧道；双侧壁导坑法；掌子面破坏；数值模拟

Study on face failure characteristics of shallow buried tunnel in Ⅴ rock

Nie Aoxiang
(Beijing General Municipal Engineering Design Research Institute Co. , Ltd. , Beijing 100082, China)

Abstract: The collapse and failure of the face is one of the major engineering problems in the construction of the shallow tunnel with soft surrounding rock. Based on this phenomenon, the failure characteristics of double side drift method were studied by means of the 3D FEM. Researches have shown that: 1) The purpose of controlling face stability is difficult to be achieved by advanced support only during the excavation of the large-section tunnel with soft surrounding rock, and the face must be reinforced to improve physical and mechanical parameters of surrounding rock in order to increase face stability. 2) During the construction of the double side drift method, the interaction caused by the excavation of each drift can not be ignored. The closer the distance is, the more obvious the interaction is. 3) When double side drift method is adopted, the excavation of the upper bench in the middle drift is liable to reach instability, so special attention should be paid to this area during construction. 4) In order to prevent the improper expansion of the failure of the face

and the connection between different destructive zones, the digging length should be limited under 1.0m and it is not suitable to excavate multiple drift at the same time.
Keywords: V rock; large-section tunnel; double side drift method; failure characteristics; numerical simulation

软基连拱景观桥设计与施工关键技术

李孟然　赵永刚
（黄河勘测规划设计研究院有限公司，河南 郑州，450003）

摘　要：多跨连拱桥在竖向荷载和温度荷载作用下会产生较大的拱脚水平推力和拱圈内力，特别是在软土地基上建造连拱桥，往往因地基变形会出现问题，如何控制或减小连拱桥拱脚水平推力和拱圈内力，是设计和施工的关键。乌海市人民路桥为内蒙古乌海市甘德尔河上的一座景观桥梁，桥梁为 11 孔跨径渐变的上承式钢筋混凝土连拱结构，桥长 173.4m，桥宽 28m。本文介绍了该桥的基本情况，设计和施工关键技术，计算分析及处理方法，希望能为类似工程提供参考。

关键词：连拱桥；景观桥；腹拱；软土地基；拱上建筑；施工

Key technology of design and construction of multi-arch landscape bridge on soft foundation

Li Mengran　Zhao Yonggang
(Yellow River Engineering Consulting Co., Ltd., Zhengzhou, Henan, 450003, China)

Abstract: Multi-span multi-arch bridge will produce greater horizontal thrust of Arch Foot and internal force of arch ring under vertical load and temperature load, especially when multi-arch bridge is built on soft soil foundation, how to control or reduce the horizontal thrust of the Arch Foot and the internal force of the arch ring is the key to design and construction. Wuhai City People's Road Bridge is a landscape bridge over the Gander River in Wuhai, Inner Mongolia. The bridge is a 11-span, tapered, deck-type reinforced concrete multi-arch structure. The length of the bridge is 173.4m and the width of the bridge is 28m. This paper introduces the basic situation, key technology of design and construction, calculation analysis and treatment method of the bridge, hoping to provide a reference for similar projects.

Keywords: multi-arch bridge; landscape bridge; spandrel arch; soft soil foundation; over-arch construction; construction

岩溶隧道大体积涌泥处治与预控技术

熊成宇　姚锐丹

（中交一公局第四工程有限公司，广西 南宁，530033）

摘　要：鱼洞Ⅰ号隧道位于贵州省余庆至凯里高速公路重安至鱼洞境内，为分离式长隧道，左幅长1374m，右幅长1379m，最大埋深173m。隧道洞身围岩为白云质灰岩，节理裂隙发育，岩体较破碎，镶嵌碎裂结构。隧道地质复杂主要不良地质为：岩溶、塌陷、瓦斯和采空区，施工安全风险高。施工场区地表见多个大小不一的小型溶洞、落水洞，地表溶沟溶槽发育，施工中出现大体积涌泥。本文针对该隧道大体积涌泥处治，从逃生通道、地表处理、洞内治理等技术措施上进行分析、总结，同时在隧道施工中采取长、短结合的地质预报方法、加强地质钻探、实施循环加深炮孔、掌子面素描等预控技术，有效地控制了涌泥突出，保证了施工安全。通过本隧道实例论证，对大体积岩溶涌泥隧道，采取大管棚结合系统小导管注浆超前支护技术，新增逃生通道，长、短结合综合地质预报、预判等技术，对保证岩溶隧道安全施工是有效的，对类似岩溶隧道施工有较好的借鉴作用。

关键词：岩溶隧道；涌泥处治；预控

Treatment and pre-control technology of massive mud in karst tunnel

Xiong Chengyu　Yao Ruidan

(The fourth engineering Co., Ltd. of CCCC first highway engineering Co., Ltd., Nanning, Guangxi 530033, China)

Abstract: Yudong No. 1 tunnel is located in the territory of Chong'an to Yudong on the Yuqing-Kaili Expressway in Guizhou Province. It is a separated long tunnel with a left width of 1374 meters, a right width of 1379 meters and a maximum buried depth of 173 meters. The surrounding rock of the tunnel body is dolomitic limestone with well-developed joints and fissures, and the rock mass is relatively broken with inlaid fragmented structures. The tunnel geology is complex and the main unfavorable geology is: karst, subsidence, gas and mined-out areas, and construction safety risks are high. There are many small karst caves and sinkholes of different sizes on the surface of the construction site. The surface karst ditch and karst grooves are developed, and a large volume of mud gushing occurs during the construction. This article analyzes and summarizes the technical measures for the treatment of the massive mud in the tunnel from the escape channel, surface treatment, and treatment in the tunnel. At the same time, the long and short geolog-

ical prediction method is adopted in the tunnel construction, the geological drilling is strengthened, and the circulation deepening is implemented. Pre-control technologies such as blast holes and face sketching effectively control the outburst of mud and ensure construction safety. Through the demonstration of this tunnel example, for large-volume karst mud-gushing tunnels, the use of large pipe shed combined with system small pipe grouting advanced support technology, newly-added escape channels, long and short combined with comprehensive geological prediction, prediction and other technologies, is helpful to ensure karst The safe construction of the tunnel is effective, and it can be used as a reference for the construction of similar karst tunnels.
Keywords: karst tunnel; mud treatment; pre-control

石灰岩隧道围岩爆破损伤区数值模拟与实测分析

杨 帆[1] 张庆明[1] 刁 吉[1] 史小雄[2] 黄 锋[2] 童小东[2]
(1. 重庆交通建设(集团)有限责任公司,重庆,401121;
2. 重庆交通大学 土木工程学院,重庆,400047)

摘 要: 山岭隧道施工通常采用光面爆破进行开挖,爆炸应力波不可避免对围岩造成动力扰动和损伤,进而影响隧道整体稳定性。论文分别进行了有限元数值模拟和现场声波测试,依托奉建高速庙宇隧道,分析了周边孔爆破作用下石灰岩地层中等效应力传播过程和围岩损伤演化机理。研究结果表明:数值模拟所得围岩损伤半径 0.9m 小于现场测试结果 1.2m 和 1.4m,综合理论计算所得 0.48m 后为 1.38m 与现场测试结果接近,这是由于现场声波测试获得的损伤区包含了静力松动圈和爆破损伤区,大于仅考虑动力扰动区的数值模拟结果。该研究可对后续的爆破工程提供数据经验和理论支撑,对光面爆破参数优化、改善爆破效果以及减少爆破对围岩的损伤有一定的指导意义。

关键词: 爆破开挖;围岩损伤;数值模拟;等效应力;声波测试

Numerical simulation and measurement analysis of blasting damage zone in limestone tunnel surrounding rock

Yang Fan[1] *Zhang Qingming*[1] *Diao Ji*[1] *Shi Xiaoxiong*[2]
Huang Feng[2] *Tong Xiaodong*[2]
(1. Chongqing Communications Construction (Group) Co., Ltd., Chongqing 401121, China;
2. School of Civil Engineering & Architecture, Chongqing Jiaotong University,
Chongqing 400074, China)

Abstract: Mountain tunnel construction usually adopts smooth blasting for excavation. The blast stress waves will inevitably destroy the dynamic disturbance and surrounding rock, and affect the overall stability of the tunnel. In this paper, with the Fengjian high-speed temple tunnel as the background, the finite element numerical simulation and the field acoustic wave test are carried out respectively., to analyze the equivalent stress propagation process in limestone formation and surrounding rock damage evolution mechanism under the action of peripheral hole blasting. The research results show that the surrounding rock damage radius of 0.9m obtained by numerical simulation is smaller than the field test results of 1.2m and 1.4m. The comprehensive theoretical calculation is 1.38m after 0.48m, which is close to the field test result. This is because the damage zone obtained by

the on-site acoustic wave test includes the static loosening circle and the blasting damage zone, which is larger than the numerical simulation result that only considers the dynamic disturbance zone. This research can provide data experience and theoretical support for subsequent blasting engineering, and has certain guiding significance for optimizing smooth blasting parameters, improving blasting effects, and reducing blasting damage to surrounding rock.
Keywords: blasting excavation; surrounding rock damage; numerical simulation; equivalent stress; sonic test

软土路基段市政道路开挖施工对浅埋地铁区间隧道结构的安全影响分析

邹淑国[1]　江　波[2]　马俊风[1]

（1. 青岛市市政工程设计研究院有限责任公司，山东 青岛 266061；
2. 青岛高新区投资开发集团有限公司，山东 青岛，266109）

摘　要：青岛市高新区田海路建设工程全线位于软土地基范围内，且部分路段处于青岛地铁8号线上方，与地铁线位重合段道路的开挖底高程与地铁拱顶的最小净距仅为7.3m，工程施工时地铁已开通运营，对地铁结构影响等级为"特级"。为避免施工过程中大面积开挖及回填引起地铁隧道结构的过大变形，工程采用堆载预压联合换填石渣的处理方法对软土地基进行加固，并采取分段跳仓施工的小步序流水作业施工方案。工程运用Midas GTS NX有限元计算软件模拟道路施工过程对下方地铁隧道结构的安全性影响，结合监测数据对预测结果进行检验，研究显示：堆载预压联合换填石渣的软基处理方法及分段跳仓施工的小步序流水作业施工方案可有效控制浅埋地铁隧道结构的变形。

关键词：超载预压联合换填石渣；分段跳仓施工；小步序流水作业；地铁隧道结构变形

Analysis on the safety impact of municipal road construction within soft soil subgrade section on shallow buried metro tunnels

Zou Shuguo[1]　*Jiang Bo*[2]　*Ma Junfeng*[1]

（1. Qingdao Municipal Engineering Design & Research Institute Co.，Ltd.，Qingdao，Shandong 266061，China；2. Qingdao High-tech Investment Group，Qingdao，Shandong 266109，China）

Abstract：The construction of Tianhai road in Qingdao high-tech zone is all within the scope of soft soil subgrade，and part of the section above the Metro Line 8，The minimum clear distance between the excavated bottom elevation of the road and the Metro Vault is only 7.3m，During the construction of the project，the subway has been in operation，and the impact on the subway structure is classified as "Super impact" . To avoid excessive deformation of subway tunnel structure caused by large-area excavation and backfill during construction，the soft soil foundation is reinforced by the method of Overloading and preloading combined with replacement of slag，Construction scheme of Small-step construction process with Sectional skip construction is adopted. MIDAS GTS NX finite element software is used to simulate the influence of the road construction process on the safety of the

subway tunnel structure under the project, the results show that the method of soft foundation treatment and the construction scheme can effectively control the deformation of subway tunnel structure.
Keywords: overloading and preloading combined with replacement of slag; sectional skip construction; small-step construction process; structural deformation of subway tunnel

大跨度曲线钢混组合梁桥设计

展丙来　钟　海　凌晓政
（中交第一公路勘察设计研究院有限公司，陕西 西安，710000）

摘　要： 互通立交中大跨曲线匝道桥采用曲线箱型钢混组合梁是一种较经济合理的结构形式。曲线箱型钢混组合梁将面临和曲线混凝土梁同样的关键问题，即合理的解决倾覆、爬移、负弯矩区的抗裂、合理水平约束等问题，设计明显比直线组合桥梁复杂。目前针对该类型桥梁的设计分析仍研究较少。论文针对互通区大跨度曲线钢混组合梁桥关键的弯扭组合受力、抗倾覆、负弯矩区抗裂、曲线梁水平约束等问题，依托49m+40m跨连续曲线组合梁工程，建立了有限元模型，分析了该类型曲梁的安全性能，探讨了负弯矩区桥面板裂缝控制措施效果，研究了门架墩水平约束合理性和混凝土桥面收缩徐变的影响，可为类似桥梁提供参考。

关键词： 桥梁工程；钢混组合梁；设计施工；结构分析；负弯矩区裂缝

Design of steel-concrete composite girder bridge with long span curve

Zhan Binglai　Zhong Hai　Ling Xiaozheng
(CCCC First Highway Consultants Co., Ltd., Xi'an, Shanxi 710000, China)

Abstract: Curved box steel composite beams will face the same key problems as curved concrete beams, which need to reasonably solve such problems as overturning, climbing, cracking resistance in negative bending moment zone, reasonable horizontal restraint, etc. The design is obviously more complex than that of linear Bridges. For each section steel composite girder bridge of large span curve combination of critical bending and twisting force, resistance to capsizing, negative bending moment OuDeKang climb beam problems such as crack, curve, paper relying on 49m+40m span continuous composite beams curve engineering, the finite element model was established, the security performance of the curved beam are analyzed, discussed the bridge panel crack control measures in the negative moment region, studies the door frame piers level constraints rationality, It can provide reference for similar bridge.

Keywords: bridge engineering; steel-concrete composite beam; design and construction; structural analysis; cracks in negative bending moment zone

考虑热位差的特长隧道斜井反井法施工通风研究

马希平

（中铁十二局集团第三工程有限公司，山西 太原，030024）

摘　要：在特长隧道的修建过程中，斜井的快速施工是影响整个工程进度的一个关键因素，当前反井法在斜井的施工中得到大量的应用及推广。而在其施工过程中，由于掌子面与交叉口热位差的存在极大地增加了通风难度。基于以上背景，本文采用数值模拟的方法，利用有限元计算软件FLUENT研究了热位差对斜井反井法施工通风的影响，研究结果表明：热位差的存在不利于洞内污染物的排除，在通风系统中产生阻力的不利影响，因此在反井法施工通风设计中，计算沿程阻力时，应考虑热位差对于洞内风流的影响。文章结合现场测试数据验证了数值计算结果的可靠性，同时考虑热位差的不利影响，对斜井反井法施工通风阻力计算公式进行了修正。研究成果在二郎山特长隧道的施工中得到了应用与验证，可为今后其他类似工程提供参考和依据。

关键词：特长隧道；反井法；热位差；数值模拟；现场测试；通风阻力

Study on ventilation of inclined shaft of extra long tunnel by reverse well method considering thermal potential difference

Ma Xiping

（China Railway 12th Bureau Group No. 3 Engineering Co.，Ltd.，
Taiyuan，Shanxi 030024，China）

Abstract：In the construction process of extra long tunnel，the rapid construction of inclined shaft is a key factor affecting the whole project progress. At present，the reverse well method has been widely used in the construction of inclined shaft. In the construction process，the existence of thermal potential difference between the tunnel face and the intersection greatly increases the difficulty of ventilation. Based on the above background，this paper uses the numerical simulation method，using the finite element calculation software FLUENT to study the influence of thermal potential difference on the ventilation of inclined shaft construction. The research results show that the existence of thermal potential difference is not conducive to the removal of pollutants in the tunnel，and it has adverse effects on the resistance in the ventilation system. Therefore，in the ventilation design of inclined shaft construction，when calculating the resistance along the way，the thermal po-

tential difference is not conducive to the removal of pollutants in the tunnel, The influence of thermal potential difference on air flow in tunnel should be considered. The reliability of the numerical results is verified by the field test data, and the calculation formula of ventilation resistance of inclined shaft reverse well method is modified considering the adverse effect of heat potential difference. The research results have been applied to the construction of Erlangshan super long tunnel, which can provide reference and basis for other similar projects in the future.
Keywords: extra long tunnel; inverted well method; thermal potential difference; numerical simulation; field test; ventilation resistance

长大下坡运梁及小半径曲线架梁综合施工技术

马希平

（中铁十二局集团第三工程有限公司，山西 太原，030024）

摘　要：桥梁预制梁运梁和架梁施工过程中，由于大型预制构件具有重量大、跨度大特点，导致特殊施工条件下难以保证预制梁质量和施工安全，尤其在长大下坡、小半径曲线以及狭窄的现场施工条件等复杂环境下给运梁和架梁带来巨大施工技术难题。本文以四湾大桥T梁的运输与安装为例进行分析，针对连续长大下坡（$L=5.8\mathrm{km}$、$i_{max}=7\%$）、小曲率半径（$R=30\mathrm{m}$）回头曲线运梁通道，提出从防倾覆控制、预制梁固定、运行控制等方面考虑的运梁施工技术；针对高墩（$H_{max}=31.2\mathrm{m}$）上的小半径曲线（$R=121\mathrm{m}$）线路、狭窄施工场地，在对两台汽车吊设备进行安全验算的基础上合理确定汽车吊架梁的机械配套选择及综合施工技术。实践表明，于长大下坡、小半径回头曲线运梁通道、小半径曲线架梁非一般施工条件下采用合理运梁措施和综合架梁施工技术，能够提高整体施工安全性，可为类似工程提供借鉴。

关键词：连续长大下坡；运梁；小半径曲线；汽车吊；架梁；机械配套

Comprehensive construction technology of beam with long heavy downgrade and girder erecting with minor radius curve

Ma Xiping

(China Railway 12th Bureau Group Co. , Ltd. , Taiyuan, Shanxi 030024, China)

Abstract: In the construction process of transporting and erecting precast beam of bridge, due to the characteristics of large weight and large span of large precast components, it is difficult to ensure the quality and construction safety of precast beam under special construction conditions, especially in the complex environment of long and large downhill, small radius curve and narrow site construction conditions. Taking the transportation and installation of T-beam of Siwan bridge as an example, aiming at the continuous long and steep downhill ($L=5.8\mathrm{km}$, $i_{max}=7\%$) and small curvature radius ($R=30\mathrm{m}$) turning curve beam transportation channel, the paper puts forward the construction technology of beam transportation from the aspects of anti overturning control, precast beam fixation and operation control; Aiming at the small radius curve ($R=121\mathrm{m}$) line and narrow construction site on High Pier ($H_{max}=31.2\mathrm{m}$), the mechanical matching selection and comprehensive construction technology of automobile hanger beam are reasonably determined on the

basis of safety checking calculation of two automobile crane equipment. The practice shows that the reasonable beam transportation measures and comprehensive beam erection construction technology can improve the overall construction safety under the unusual construction conditions of long and large downhill, small radius curved beam transportation channel and small radius curved beam erection, which can provide reference for similar projects.
Keywords: continuous long heavy downgrade; beam; minor radius curve; truck crane; girder erecting; mechanical mating

堆载作用下深埋软土地基桥桩负摩阻力分析

展丙来　钟　海
（中交第一公路勘察设计研究院有限公司，陕西 西安，710000）

摘　要：对于软土层埋深大于20m层厚10～30m的特殊场地，桥梁桩基设计不可避免地需要解决负摩阻力问题。目前，国内外关于分析软土地区桩基负摩阻力对桩基计算的影响的研究主要集中于浅埋软弱土地基、欠固结软土地基、回填土地基等方面，但是对于软土层埋深大于20m层厚10～30m的特殊场地桩基负摩阻力问题研究分析鲜见报道。依托背景工程及地质资料，论文阐述了场地的特殊性、负摩阻力发展的机理，定性分析了负摩阻力影响范围，根据现行桥梁规范推荐的计算算法，对典型段落进行了负摩阻力的计算，发现负摩阻力的影响巨大，需要在桥梁设计中充分重视以保证桥梁结构安全，结果可供类似工程参考。

关键词：桥梁工程；设计；桩基；负摩阻力

Analysis of negative friction resistance of bridge piles in deep soft soil under surcharge

Zhan Binglai　Zhong Hai
(CCCC First Highway Consulatants Co., Ltd., Xi′an, Shanxi 710000, China)

Abstract: For the special site where the buried depth of soft soil layer is greater than 20m and the layer thickness is 10—30m, the problem of negative friction resistance is inevitably needed to be solved in the design of bridge pile foundation. Relying on the background of engineering and geological data, the paper expounds the particularity of space, the development mechanism of negative skin friction resistance, the qualitative analysis of the impact scope, the negative skin friction resistance bridge according to the current specification recommended calculation algorithm, a typical paragraph has carried on the calculation of negative skin friction resistance, found that the influence of negative skin friction resistance is huge, need full attention in bridge design in order to ensure the safety of bridge structure. The results can be used as reference for similar projects.

Keywords: bridge engineering; design; pile foundation; the negative friction resistance

广西拱桥建造技术的发展与创新

杜海龙　莫昀锦　马博彧
（广西路桥工程集团有限公司，广西 南宁，530200）

摘　要：为了解广西拱桥建造技术的发展情况，本文分析了采用支架法、无支架缆索吊装技术、钢丝绳斜拉扣挂松索成拱技术和千斤顶钢绞线斜拉扣挂技术建造的拱桥的建设和技术创新情况，并展望了广西拱桥今后的技术创新方向和发展趋势。分析结果表明：无支架缆索吊装技术有效地解决了不搭设支架建设拱桥的难题，在此基础上进一步发展的千斤顶钢绞线斜拉扣挂合龙后松索施工技术有效地解决了大跨径拱桥多节段吊装的问题。得益于缆索吊装斜拉扣挂技术的不断发展和创新，拱桥有望在跨径方面取得进一步突破。
关键词：桥梁工程；拱桥；施工；缆索吊装；斜拉扣挂

The development and innovation of construction technology of arch bridge in Guangxi

Du Hailong　Mo Yunjin　Ma Boyu
（Guangxi Road and Bridge Engineering Group Co.，Ltd.，
Nanning，Guangxi 530200，China）

Abstract: In order to understand the development of construction technology of arch bridge in Guangxi, this paper analyzes the construction methods and technological innovation in arch bridge building by frame support method, cable hoisting technology, steel wires for cable-stayed buckling technology and strand jacks for cable-stayed buckling technology and focuses on the direction of technological innovation and development trends for arch bridge in Guangxi in the future. The analysis result indicates that by adopting cable hoisting technology, which effectively eliminated frame support in arch bridge construction, and based on that the further improved strand jack for cable-stayed buckling effectively solved the problem of multiple segments hoisting for long span arch bridge after its closure. Taking advantage of constant development and innovation of cable hoisting and cable-stayed buckling technology, the arch bridge is expected to make a breakthrough at its span.
Keywords: bridge engineering; arch bridge; construction; cable hoisting; cable-stayed buckling

基于双向扭曲连续钢箱梁倒装制作与正装拼装的施工技术

孙晓伟　窦市鹏　史　伟　尹建鲁　赵　阳
（中建八局钢结构工程有限公司，上海，201304）

摘　要：二环北路及东西延伸段智慧快速路水上钢箱梁为空间双向扭曲结构，采用正交异性桥面结构。建立BIM模型，将模型进行纵向分段和横向分片，提取分段和隔板处的坐标控制点，进行胎架设计。采用钢箱梁倒装制作和二次正装拼装技术，能够确保钢箱梁几何精度、焊接质量，施工安全可靠，能够提高安装效率，确保钢箱梁和线型符合设计和规范要求。
关键词：水上钢箱梁；正交异性；分段；坐标；胎架；倒装；正装

Construction technology of flip fabrication and formalization of continuous steel box girder based on two-way twist

Sun Xiaowei　Dou shipeng　Shi Wei　Yin Jianlu　Zhao Yang
(China Construction Eighth Engineering Bureau Co., Ltd., Shanghai 201304, China)

Abstract: The steel box girder of the North Second Ring Road and the east-west Extension Section of the smart expressway is a spatial two-way twisted structure with orthotropic deck structure. The BIM model was built, and the model was divided into vertical and horizontal sections, and the coordinate control points at the sections and partitions were extracted to design the frame. The technology of flip-flop fabrication and double normal assembling of steel box girder can ensure the geometric precision, welding quality, construction safety and reliability, increase installation efficiency, and ensure that the steel box girder and line shape meet the requirements of design and specification.
Keywords: steel box girder on water; orthotropic; segmentation; coordinate; frame; inversion; formal

西南地区隧道富水异构区域超前探测模型及方法

刘常昊[1]　杨志全[1]　郑万波[1,2]　吴燕清[3]　史耀轩[2]
（1. 昆明理工大学，云南 昆明，650504；2. 昆明理工大学，云南 昆明，650500；
3. 重庆大学，重庆，400030）

摘　要：为解决西南地区隧道挖掘过程中因富水异构区产生的突涌水现象而制约施工进度的难题，提出了一种基于地质雷达的富水异构区域结构模型并引入涌水量预测公式的方法，通过综合采用 MATLAB 正演仿真模拟、数学建模、公式改进等方法，建立富水异构区域结构模型以及改进隧道初期最大涌水量公式，为隧道挖掘初期突涌水预测提供了理论依据，弥补了传统地质雷达对富水区域探测，仅能粗略估计富水区域位置而不能计算其最大初期涌水量的不足，并结合云南玉磨铁路景寨隧道工程实例验证，同传统大岛洋志公式比较，提升精度 66.3%，预测效果较好，为西南地区隧道富水异构体探测提供了一种科学的参考依据和技术支撑，可为类似工程提供借鉴。

关键词：隧道工程；隧道富水；地质超前预报；模型试验；涌水量预测

Research and application of structural characteristics of tunnel water rich area based on ground Penetrating radar

Liu Changhao[1]　*Yang Zhiquan*　*Zheng Wanbo*[1,2]　*Wu Yanqing*[3]　*Shi Yaoxuan*[2]
(1, Kunming University of science and technology, Kunming, Yunnan 650504, China;
2. , Kunming University of Science and Technology, Kunming , Yunnan 650500, China;
3. Chongqing University, Chongqing 400030, China)

Abstract: In order to solve the problem of restricting the construction progress due to the water inrush phenomenon in the process of tunnel excavation in Southwest China, a structural model of water rich heterogeneous area based on ground penetrating radar is proposed, and the method of water inflow prediction formula is introduced. Through the comprehensive use of MATLAB forward simulation, mathematical modeling, formula improvement and other methods, This paper establishes the heterogeneous regional structure model of rich water and improves the formula of maximum water inflow at the initial stage of tunnel excavation, which provides a theoretical basis for the prediction of water inrush at the initial stage of tunnel excavation. It makes up for the deficiency that the traditional ground penetrating radar can only roughly estimate the location of rich water area, but can not calculate the maximum initial water inflow, Compared with the traditional Oshima Yo-

shi formula, the accuracy of the simultaneous interpreting method is 66.3% and the prediction effect is better. It provides a scientific reference and technical support for the exploration of water rich isomers in the southwest area of the tunnel, and can provide reference for similar projects.
Keywords: tunnel engineering; the tunnel is rich in water; advanced geological prediction; model test; prediction of water inflow

自锚式悬索桥体系转换“M”型吊杆张拉法

秦建刚　何龙虎　姜云晖　徐朝政
（中建三局城市投资运营有限公司，湖北 武汉，430074）

摘　要： 针对自锚式悬索桥施工中体系转换的关键问题——吊杆张拉，首次提出了一种“M”型吊杆张拉法，即将跨中和1/4中跨处吊杆同时张拉，控制主缆“M”变形的关键点，一方面可以使主缆变形较快从空缆向成桥线形逼近，另一方面1/4中跨处基本位于钢梁最大临时墩跨径的跨中附近，故张拉此处吊杆可以加速使主梁脱离临时墩。随着体系转换的进行，适当的时机再进行二次或三次张拉该处吊杆，可以使其他吊杆较快张拉至成桥无应力长度。该方法相比于传统的张拉方法可较大程度减少吊杆张拉次数和接长杆用量，提高体系转换的效率，在双塔自锚式悬索桥体系转换中具有通用性。

关键词： 自锚式悬索桥；体系转换；“M”型；吊杆张拉；有限元法

Type M suspender tensioning method in system transformation of self anchored suspension bridge

Qin Jiangang　He Longhu　Jiang Yunhui　Xu Chaozheng
(City Investment and Operation Co., Ltd. of China Construction Third Engineering Bureau, Wuhan, Hubei, 430074, China)

Abstract: According to the key problem of system transformation of self anchored suspension bridge in construction—tensioning suspender, type M suspender tensioning method is first put forward, tensioning the suspenders in the middle and 1/4 midspan at the same time, controlling the key points of type "M" deformation of main cable. On the one hand, the tension make the main cable transform from the free cable state to the finished bridge state fast, on the other hand, 1/4 midspan is nearly located in the middle of the span of the largest temporary pier, so tensioning suspenders here can make the girder divorce from temporary piers quickly. With the system transformation, tensioning the suspenders two or three times in the right time, which can make the other suspenders quickly tension to the unstressed length in finished bridge state. Comparing with the traditional tension method, the method can greatly reduce the tension times of suspenders and the amount of extension bar, improve the efficiency of system transformation. It's universal for the self anchored suspension bridge with double tower.

Keywords: self anchored bridge; system transformation; type M; tensioning suspender; finite element method

公轨两用空间缆索悬索桥动力特性与车致振动研究

周　涛　王　鹏　闫海青
（长江勘测规划设计研究有限责任公司，湖北 武汉，430010）

摘　要：同时承受公路、地铁荷载的空间缆索悬索桥，车桥振动问题突出，有必要对其动力特性与车致振动进行研究。基于某跨度为330m的空间缆索悬索桥，首先通过调整两塔顶I、P点横向间距，研究索面由内敛式到平行索面，再到外张式变化时，结构动力特性的变化情况。发现索面倾角越大，结构横漂频率越大，得出空间缆索可以提高悬索桥的横向刚度。其次采用移动质量模拟车辆过桥，在ANSYS中进行瞬态非线性时程分析，分析车辆激励下结构的动力响应。结果表明车辆速度和质量越大，结构响应越大；汽车行车间距越大，结构响应越小，当汽车行车间距为22.5m时，车辆激励频率与结构三阶竖弯振动频率接近，激发结构共振现象，位移和加速度响应幅值明显增大；双向列车过桥时，结构动力响应比单向行车时大很多。可以通过限载、规范汽车行车间距、避免双向列车同时过桥等措施减小结构动力响应。

关键词：空间缆索；动力特性；车致振动；瞬态非线性时程分析

Dynamic characteristics and vehicle-induced vibration of suspension bridge with spatial cables

Zhou Tao　Wang Peng　Yan Haiqing
(Changjiang Institute of Survey, Planning, Design and Research,
Wuhan, Hubei 430010, China)

Abstract: For suspension bridge with spatial cables, the vehicle-induced vibration is significant, when subjected to road and subway loading. It is therefore necessary to study the dynamic characteristics as well as the vehicle-induced vibration of such structures. In the present work, a spatial-cable suspension bridge with a span of 330m is taken for a case study. First, by changing the lateral distance between the I. P. points, the cable surface changes from inward to parallel, and then to outward, yielding in different models for the analysis of dynamic characteristics. The results show that the natural frequency of the lateral mode becomes higher with the increasing angle of the cable surface, which indicates that spatial cables are helpful to improve the lateral stiffness of suspension bridge. Next, the vehicle travelling through the bridge, simulated by a moving mass, is studied by the transient nonlinear time-history analysis in ANSYS. The results show that the structural

vibration gets more intensive with the increasing speed or mass of the vehicles. On the other hand, the structural vibration gets weaker when increasing the distance between the vehicles. Particularly, when such distance is 22.5m, the vehicle excitation, having the frequency close to the third natural frequency of the structural vertical bending, induces the resonance, which significantly amplifies the displacement and acceleration response. Additionally, the dynamic response of the structure is much larger when two way trains passing through. From this study, it can be concluded that the structural vibration can be reduced by several measures, e. g. , controlling the weight of vehicle or the distance between vehicles, avoiding two way trains passing through simultaneously.
Keywords: spatial cables; dynamic characteristics; vehicle-induced vibration; nonlinear time-history analysis

桥梁深水岩石河床双壁钢围堰非爆破开挖施工技术

王 勇 董传洲 赵研华
（中建三局第三建设工程有限责任公司，湖北 武汉，430074）

摘 要： 随着城市交通的发展需求，不可避免会出现紧挨既有的桥梁或重要建筑增加桥梁，如遇河床无覆盖层，在桥梁埋入式基础不能采取爆破开挖施工时，桥梁基础施工将会是一个难题。重庆轨道交通九号线嘉华轨道专用桥跨越嘉陵江，距原有嘉华大桥 80m，河床为无覆盖层砂岩，必须采取非爆破开挖水中 P5 墩。采用先机械开挖双壁钢围堰基槽，下放围堰，浇筑围堰内外混凝土将围堰嵌固在河床内，在围堰内开挖承台基坑和施工水下结构。嘉陵江水位变化大，施工期间水位变化 13～27m，水流流速高 1.2～3.5m，该围堰未出现渗漏和变形。该围堰施工方法可靠，可供其他类似条件桥梁施工借鉴。

关键词： 桥梁基础；非爆破开挖；双壁钢围堰；岩石河床；深水

Non blasting excavation construction technology for double wall steel cofferdam of bridge deep water rock riverbed

Wang Yong Dong Chuanzhou Zhao Yanhua
(The Third Construction Co., Ltd. of China Construction Third Engineering Bureau, Wuhan, Hubei, 430074, China)

Abstract: with the development of urban traffic demand, it is inevitable to increase bridges close to existing bridges or important buildings. If there is no overburden in the riverbed, the bridge foundation construction will be a difficult problem when the bridge embedded foundation can not adopt blasting excavation construction. The Jiahua rail bridge of Chongqing Rail Transit Line 9 crosses Jialing River, 80m away from the original Jiahua bridge. The riverbed is sandstone without overburden, so the P5 pier in water must be excavated without blasting. Firstly, the foundation trench of double wall steel cofferdam is mechanically excavated, the cofferdam is lowered, the concrete inside and outside the cofferdam is poured, the cofferdam is embedded in the river bed, the foundation pit of pile cap is excavated in the cofferdam and the underwater structure is constructed. The water level of Jialing River changes greatly. During the construction period, the water level changes

by 13-27m, and the flow velocity is 1. 2-3. 5m. The cofferdam has no leakage and deformation. The cofferdam construction method is reliable, which can be used for reference for other similar bridge construction.
Keywords: bridge foundation; non blasting excavation; double wall steel cofferdam; rock riverbed; deep water

基于GRU神经网络的都四轨道交通映秀一号隧道瓦斯浓度序列预测

丁力生[1]　赖永标[1]　杨　扬[1]　赖祥威[2,3]
郑万波[2,3]　杨黎明[1]　王　飞[1]　周　辉[1]　冉啟华[4]
（1. 中建铁路投资建设集团有限公司 北京，102601；2. 昆明理工大学 理学院，云南 昆明，650500；3. 昆明理工大学 数据科学研究中心，云南 昆明，650500；4. 云南卫士盾科技有限公司，云南 昆明，650500）

摘　要：为解决具有时序性和非线性的瓦斯浓度序列预测误差较大的问题，利用门控循环单元神经网络对其预测。该算法对数据集进行预处理，接着引入更新门和重置门，设计出门控循环单元神经网络瓦斯浓度序列预测算法结构，以误差损失最小化为目标，得到训练模型完成瓦斯浓度预测。以都四轨道交通映秀一号隧道瓦斯监控数据为实例，利用该算法预测瓦斯浓度，该算法预测得到的最小均方根误差为5.421%，最小平均绝对误差为1.391%，并与卷积神经网络、循环神经网络和多层感知机进行对比，实验揭示，该算法可提高瓦斯浓度预测精度。

关键词：门控循环单元；卷积神经网络；循环神经网络；多层感知机；瓦斯浓度预测

Prediction of gas concentration sequence inYingxiu No. 1 Tunnel of dusi metro based on GRU neural network

Ding Lisheng[1]　*Lai Yongbiao*[1]　*Yang Yang*[1]　*Lai Xiangwei*[2,3]
Zheng Wanbo[2,3]　*Yang Liming*[1]　*Wang Fei*[1]　*Zhou Hui*[1]　*Ran Qihua*[4]
（1. China Construction Railway Investment and Construction Group Co., Ltd., Beijing 102601, China; 2. Faculty of Science, Kunming University of Science and Technology, Kunming, Yunnan 650500, China; 3. Data Science Research Center, Kunming University of Science and Technology, Kunming, Yunnan 650500, China; 4. YunNan Wisdom Sience Technology Co., Ltd., Kunming, Yunnan 650500, China）

Abstract: In order to solve the problem of large error in the prediction of gas concentration series with time sequence and non-linearity, the gated cyclic unit neural network is used to predict the gas concentration series. The algorithm preprocesses the data set, and then introduces the update gate and reset gate to design the algorithm structure of gas concentration sequence prediction based on the outdoor control loop unit neural network. With the

goal of minimizing the error loss, a training model is obtained to complete the gas concentration prediction. In all four metro tunnel Yingxiu a gas monitoring data as an example, using the algorithm to predict the gas concentration, and the algorithm to predict the minimum root mean square error was 5. 421%, the minimum average absolute error is 1. 391%, and with convolution neural network, circulation and multilayer perceptron neural network, the experiment has revealed that the algorithm can improve the gas concentration prediction accuracy.
Keywords: gated circulation unit (GRU); convolutional neural network (CNN); recurrent neural network (RNN); multilayer perceptron; gas concentration prediction

宽桥面单片拱梁拱组合体系桥吊点横梁分析

朱克兆　王　鹏　尤　岭
（长江勘测规划设计研究有限责任公司，湖北 武汉，430010）

摘　要：宽桥面单片拱梁拱组合体系桥可以提高主梁刚度，有效改善大跨度梁桥跨中下挠以及开裂问题，同时可以减小主梁结构的建筑高度，但是吊杆力的影响，使得横向计算较为复杂。本文结合规范，提出采用横框计算模型对吊点横梁的横向受力进行分析，并建立对应实体分析模型对横框计算结果进行验证。通过研究发现横框模型中腹板处采用刚性支撑约束的横框计算分析所得的拉应力小于按照弹性支撑约束所得结果，腹板在荷载作用下存在一定的竖向弹性变形，选用弹性支撑约束的横框模型更加准确；随着横梁高度的增加，横梁顶底缘的应力均有一定的减小，通过对比分析，采用 1.7m 高横梁承担结构横向受力是安全的；通过实体计算分析可以验证，横框计算模型所得结果较为准确且偏安全，可以用来进行类似结构的横向受力分析。

关键词：梁拱组合；横梁计算；强梁弱拱；横框分析；实体分析

Analysis of suspension point crossbeam of beam arch composite bridge with wide deck and single arch

Zhu Kezhao　Wang Peng　You Ling
(Changjiang Institute of Survey Planning, Design and Research,
Wuhan, Hubei 430010, China)

Abstract: The single arch beam arch composite bridge with wide deck can improve the stiffness of main beam, effectively improve the midspan deflection and cracking of long-span beam bridge, and reduce the building height of main beam structure. However, the influence of Suspender Force makes the transverse calculation more complex. In this paper, combined with the specification, the transverse frame calculation model is proposed to analyze the transverse force of the beam at the lifting point, and the corresponding entity analysis model is established to verify the calculation results of the transverse frame. Through the research, it is found that the tensile stress of the transverse frame model with rigid support constraint at the web plate is less than that of the elastic support constraint, and the web plate has a certain vertical elastic deformation under the load, so the elastic support constraint model is more accurate; with the increase of the beam height, the stress at the top and bottom edge of the beam decreases to a certain extent The results show that it is safe to use 1.7m high beam to bear the transverse force of the structure; through the

entity calculation and analysis, it can be verified that the results of the cross frame calculation model are more accurate and safe, which can be used for the transverse force analysis of similar structures.
Keywords: beam arch combination; crossbeam calculation; strong beam and weak arch; transom analysis; entity analysis

大跨径玻璃悬索桥复杂山岭地区隧道锚施工技术

梁隆祥　齐　帆　李志杰　翟　跃　任　明
（中国建筑第六工程局有限公司，天津，300450）

摘　要：浙江省越龙山悬崖玻璃桥是一座主跨 710m 的地锚式空间索面悬索桥，隧道锚长 57.335m，锚塞体长 16m，与水平线倾角达 29°，是同类型玻璃悬索桥中坡度最大、最长隧道式锚碇。为保证坚硬围岩下隧道锚顺利开挖，加快隧道锚施工进度，对传统隧道锚施工方法进行调查研究，研究了一种隧道锚施工技术，该技术采用两台阶开挖法，小剂量光面爆破，严格控制药量，优化起爆网络，保证围岩断面尺寸的精准性；锚固系统定位支架采用洞内分段法安装；锚塞体大体积混凝土施工采用水平分层浇筑，分层埋设冷却水管降低温差，保护层设置防裂钢筋网，并采用智能温控系统随时监控温度；锚固系统锚杆采用自制三角架运输车运输至洞口，通过橡胶滑轮组运输至洞内；采用洞顶的手拉葫芦四点吊装法进行锚杆的吊装。通过采取以上技术措施顺利完成了隧道锚施工，大大缩短了工期，有效节约了成本。

关键词：悬索桥；隧道锚；锚固系统定位支架；锚固系统锚杆；大体积混凝土；施工技术

Construction technology of tunnel anchor for long span glass suspension bridge in complex mountain area

Liang Longxiang　Qi Fan　Li Zhijie　Zhai Yue　Ren Ming
(China Construction Sixth Engineering Bureau Co., Ltd., Tianjin 300450, China)

Abstract: Yuelongshan cliff glass bridge in Zhejiang Province is a ground anchored spatial cable plane suspension bridge with a main span of 700m. The length of tunnel anchor is 57.335m, the length of anchor plug is 16m, and the inclination angle with the horizontal line is 29°, It is the longest tunnel anchorage with the largest slope in the same type of glass suspension bridge. In order to ensure the smooth excavation of tunnel anchor under hard surrounding rock and speed up the construction progress of tunnel anchor, the traditional construction method of tunnel anchor is investigated and studied, and a tunnel anchor construction technology is studied. The technology adopts two-step excavation method, small dose smooth blasting, strict control of explosive quantity, optimization of blasting network, and guarantee the accuracy of surrounding rock section size; The positioning support of the anchoring system is installed by segment method in the tunnel; The mass concrete construction of anchor plug body adopts horizontal layered pouring, laying cooling

water pipes in layers to reduce the temperature difference, setting anti crack steel mesh in the protective layer, and adopting intelligent temperature control system to monitor the temperature at any time; The anchor rod of the anchoring system is transported to the hole by self-made tripod transport vehicle and transported to the hole by rubber pulley block; The four point hoisting method of chain block on the top of the tunnel is used to hoist the anchor rod. By adopting the above technical measures, the tunnel anchor construction is successfully completed, which greatly shortens the construction period and effectively saves the cost.
Keywords: suspension bridge; tunnel anchor; anchor system positioning bracket; anchor bolt of anchoring system; mass concrete; construction technique

大跨径玻璃悬索桥重力式锚碇锚固系统施工技术

梁隆祥　齐　帆　李志杰　翟　跃　任　明
（中国建筑第六工程局有限公司，天津，300450）

摘　要： 浙江省越龙山悬崖玻璃桥是一座主跨710m的地锚式空间索面悬索桥，为提高重力式锚碇主缆型钢锚碇系统的安装精度，首先建立型钢锚固系统和定位支架整模型，通过型钢锚固系统和定位支架的设计计算，优化锚固系统定位支架设计，并研究了一种重力式锚碇锚固系统施工技术。该技术中将锚固系统定位支架分为基架＋桁架整体吊装安装方法，大大提高了锚固系统定位支架稳定性和刚度，降低了型钢锚杆定位难度，有效提高了型钢锚杆定位的精准性。

关键词： 悬索桥；重力锚；锚固系统定位支架；型钢锚杆；施工技术

Construction technology of gravity anchor for long span glass suspension bridge in complex mountain area

Liang Longxiang　Qi Fan　Li Zhijie　Zhai Yue　Ren Ming
(China Construction Sixth Engineering Bureau Co. , Ltd. , Tianjin 300450, China)

Abstract: Yuelongshan cliff glass bridge in Zhejiang Province is a ground anchored spatial cable plane suspension bridge with a main span of 710m. In order to improve the installation accuracy of gravity anchor main cable section steel anchorage system, the whole model of section steel anchorage system and positioning support is established firstly. Through the design calculation of section steel anchorage system and positioning support, the positioning support design of anchorage system is optimized, And the construction technology of a gravity anchorage system is studied. In this technology, the anchor system positioning bracket is divided into the base frame ＋ truss integral hoisting installation method, which greatly improves the stability and stiffness of the anchor system positioning bracket, reduces the positioning difficulty of the steel anchor, and effectively improves the positioning accuracy of the steel anchor.

Keywords: suspension bridge; gravity anchor; anchor system positioning bracket; section steel bolt; construction technique

小净距近接既有线高铁隧道施工技术研究

张　磊　矫　健　代广伟　丁博韬
（中建三局集团有限公司东北分公司，辽宁 沈阳，110000）

摘　要：为研究小净距并行、上跨近接既有线空间复杂隧道施工技术，根据围岩等级和隧道空间净距，通过采用控制爆破配合机械隧道施工方法，选取新建孙家沟 1、2 号隧道施工代表段进行爆破模拟试验以及通过施工观察（洞内、外）、监测手段（拱顶下沉、净空变化、结构内力）的实施，优化确定了控制爆破配合机械开挖、弱爆破修边工艺参数，并通过在既有线隧道内安装铁路设备监控元件，对既有线隧道轨道位移、隧道水平收敛、结构内力、新建隧道爆破振动等进行实时监控，分析隧道结构安全性能及轨道的工作情况以及开挖卸载作用机理，对既有线隧道进行安全性评价，指导新建隧道的施工。研究表明：遵循隧道工程与监测应用的研究规律，模拟试验优化后的控制爆破配合机械隧道施工工艺参数以及通过无线信息传输系统监控量测数据对新建隧道的施工参数的指导修正，既保证了既有线隧道的结构安全，同时确保了工期节点，实现了既有线隧道安全和新建隧道进度和谐统一。

关键词：小净距；孙家沟 1、2 号隧道；既有线隧道；爆破模拟试验；工艺参数；监控量测；安全评价

Study on construction technology of high speed railway tunnel with small clear distance and near existing line

ZhangLei　Jiao Jian　Dai Guangwei　Ding Botao
(Northeast Branch of China Construction Third Engineering Bureau Group Co., Ltd., Shenyang, Liaoning, 110000, China)

Abstract: In order to study the construction technology of complex space tunnel with small clear distance and close to the existing line, according to the surrounding rock grade and the clear distance of the tunnel space, through the use of controlled blasting combined with mechanical tunnel construction method, the blasting simulation test was carried out in the representative sections of the construction of the new Sunjiagou No. 1 and No. 2 tunnels. Through the implementation of construction observation (inside and outside of the tunnel) and monitoring means (vault subsidence, clearance change, structural internal force), the process parameters of controlled blasting combined with mechanical excavation and weak blasting trimming were optimized and determined. Through the installation of railway equipment monitoring components in the existing tunnel, the track displacement, horizontal

convergence of the tunnel, structural internal force and blasting vibration of the new tunnel were monitored in real time. The safety performance of the tunnel structure, the working condition of the track and the mechanism of excavation and unloading were analyzed. The research shows that following the research law of tunnel engineering and monitoring application, the construction parameters of the controlled blasting combined with the mechanical tunnel after the optimization of the simulation test and the guidance and correction of the construction parameters of the new tunnel through the monitoring data of the wireless information transmission system ensure the structural safety of the existing tunnel, and ensure the construction period node, and realize the harmony and unity of the safety of the existing tunnel and the progress of the new tunnel.
Keywords: small net distance; Sunjiagou tunnel 1 and 2; existing line tunnels; explosion simulation test; process parameters; monitoring measurement; safety evaluation

双柱式重力式桥墩结构大桥拆除爆破控制技术

吴 鹏　付承涛　李卫群　袁泽洲
（中国建筑第五工程局有限公司，湖南 长沙，410000；
江西省高端爆破工程有限公司，江西 吉安，343000）

摘　要： 待拆大桥为长征第一渡，具有红色意义，经技术鉴定为五类危桥，需爆破拆除。大桥无建桥资料，河床以下结构不明，混凝土强度及布筋不明，因汛期即至，工期要求紧，爆破拆除难度较大。为确保爆破时对周边构建筑物无影响，根据后期的勘探，通过采取合理的爆破部位、单耗、爆破延时及安全防护等技术措施，使爆破拆除达到了预期效果。

关键词： 结构；工期；勘探；技术措施

Demolition and blasting control technology of double-column and gravity bridge piers

Wu Peng　Fu Chengtao　Li Weiqun　Yuan Zezhou
(China Construction Fifth Engineering Division Co. , Ltd. , Changsha 410000, China;
Jiangxi High - end B lasting Engineering Co. , Ltd. , Ji'an, Jiangxi 343000, China)

Abstract: The bridge to be dismantled is the first crossing of the Long March , of red significance , technically identified as five types of dangerous Bridges, which needs to be demolished . There is no bridge construction data of the bridge, the structure below the riverbed is unknown, and the concrete strength and reinforcement are unknown. Due to the flood season, the construction period is tight, and it is difficult for blasting and demolition. In order to ensure that there is no impact on the surrounding structures during blasting, according to the later exploration, reasonable technical measures such as blasting parts, single consumption, blasting delay and safety protection are adopted to achieve the expected effect.

Keywords: structure; duration; exploration; technical measures

隧洞高地应力复核及评价方法应用研究

冉　立[1]　周权峰[1]　彭文彬[1]　裴开元[1]　周　昊[2]　高桂云[2]
(1. 中海建筑有限公司　广东 深圳，518057；
2. 应急管理部国家自然灾害防治研究院，北京，100085)

摘　要： 长大、深埋隧道围岩环境易于遭受高地应力作用，隧洞内地应力测量及准确的高地应力评价可以为隧道设计和施工方案变更及预算调整等提供有力依据和数据支撑。本文采用隧道开挖后隧洞内水压致裂地应力测量方法，获得隧道当前埋深位置的真实原地应力值，通过取样测试该埋深岩石强度并利用 Hoek-Brown 强度理论对工程岩体强度参数进行估算，然后采用基于 Hoek-Brown 强度理论的应力状态评价准则对高地应力状态进行评价，并结合开挖过程中围岩变形破坏情况，可以有效地复核及评价隧道围岩应力状态。桃子垭隧道内某钻孔的地应力复核及评价结果显示，应力量值与勘察阶段地应力测试及预测结果基本一致，该隧道段洞身部位的应力状态为中等应力状态，很好地解释了开挖过程中的破坏现象，验证了本文方法的快速、有效性，为其他类似工程的应力复核及评价提供参考。

关键词： 桃子垭隧道；应力测量；应力复核；高地应力；应力评价

Research on high in-situ stress recheck and evaluation method for tunnel and its application

Ran Li[1]　*Zhou Quanfeng*[1]　*Peng Wenbin*[1]　*Pei Kaiyuan*[1]　*Zhou Hao*[2]　*Gao Guiyun*[2]
(1. China Overseas Construction Limited, Shenzhen, Guangdong 518057, China;
2. National Institute of Natural Hazards, MEMC, Beijing 100085, China)

Abstract: Measurements and accurate evaluation of high in-situ stress in super-long and deep buried tunnels which prone to high in-situ stress could provide a powerful basis and data support for the design and construction scheme change and budget adjustment during tunnel construction. We measured the in-situ stress in tunnel after excavation by using hydraulic fracturing in-situ stress measurements to obtain the real in-situ stress. The rock mass strength parameters were estimated using the Hoek-Brown strength criterion based on the rock core compressive strength test results. Then we evaluated the in-situ stress state by using a strength-stress ratio method based on Hoek-Brown criterion. And combined with the deformation and failure of surrounding rock mass in the process of excavation, the in-situ stress state of tunnel surrounding rock can be effectively checked and evaluated. The results of stress measurements and evaluation of one borehole in Taoziya Tun-

nel show that the stress magnitudes measured during excavation are in consistent with that measured and predicted in surveying and designing stages. The in-situ stress state in this tunnel section is medium stress state, which well explains the failure phenomenon in the excavation process. This verifies the rapidity and effectiveness of the method presented in this paper, and provides a reference for stress recheck and evaluation in other similar projects.
Keywords: Taoziya tunnel; in-situ stress measurement; stress recheck; high in-situ stress; stress evaluation

浅谈高速铁路隧道衬砌混凝土智能化养护技术

黄国富
（中国建筑第五工程局有限公司，湖南 长沙，410004）

摘　要：以某高铁隧道实际工程为背景，针对隧道衬砌混凝土表观病害问题开展了现场调查，统计分析了主要病害类型与特征，并分别从施工工艺、原材料控制等角度探讨了造成衬砌表观病害的基本原因，结论指出：造成高铁隧道衬砌混凝土表观病害的关键成因之一为养护水温与混凝土内部水化热的梯度差过大。进一步，结合工程实践，研究提出了温水养护方案并开发了智能养护台车，形成了降低混凝土衬砌表观病害的智能养护技术。实践应用表明：相比于传统养护技术，智能养护技术可降低隧道衬砌表观病害发生率90%，控制效果明显。

关键词：隧道衬砌；质量；智能化养护

Discussion on intelligent curing technology of concrete lining of high-speed railway tunnel

Huang Guofu
(China Construction Fifth Engineering Division Crop. , Ltd. ,
Changsha, Hunan 410004, China)

Abstract: Based on an actual project of a high-speed railway tunnel, a field investigation was carried out on the apparent damage of concrete lining of the tunnel. The main types and characteristics of the damage were statistically analyzed, and the basic reasons for the apparent damage of lining were discussed from the perspectives of construction technology and raw material control, etc. The conclusions were as follows: One of the key causes of the apparent damage of the concrete lining of high-speed railway tunnel is the large gradient difference between the curing water temperature and the hydration heat inside the concrete. Furthermore, combining with the engineering practice, a warm water curing scheme was put forward and an intelligent curing trolley was developed, which formed an intelligent curing technology to reduce the apparent damage of concrete lining. The practical application shows that compared with the traditional curing technology, the intelligent curing technology can reduce the apparent damage rate of tunnel lining by 90%, and the control effect is obvious.

Keywords: tunnel lining; quality; intelligent maintenance

超高性能混凝土(UHPC)箱型拱桥关键技术

韩　玉　翁贻令　解威威

（广西路桥工程集团有限公司，广西 南宁，530200）

摘　要： 超高性能混凝土（UHPC）具有超高的抗压强度，与受压为主的拱桥结构结合，可大幅减轻拱肋自重，有望突破拱桥跨径。为探索新型大跨拱桥的建造技术，设计建造了国内首座 UHPC 箱型拱桥。通过对云南昌保高速一座小跨径跨线桥设计、建造、试验，研究经济型 UHPC 制备与 UHPC 拱桥建设关键技术。研究结果表明：采用 UHPC 建造拱桥可减小截面厚度，大幅减轻拱肋自重；采用机制砂制备 UHPC 不仅可满足桥梁应用需求，还可降低材料成本；桥梁各项试验结果显示，UHPC 拱桥受力性能良好。该桥的成功建造，充分显示了 UHPC 在拱桥中的应用潜力，为后续的大跨径拱桥建造提供技术积累。

关键词： 桥梁工程；箱型拱桥；车行天桥；超高性能混凝土（UHPC）；机制砂

Key Technology of Ultra High Performance Concrete (UHPC) Box Arch Bridge

Han Yu　Weng Yiling　Xie Weiwei

(Guangxi Road and Bridge Engineering Group Co., Ltd., Nanning, Cuangxi 530200, China)

Abstract: Ultra-high performance concrete (UHPC) has ultra-high compressive strength. The combination of UHPC and the compression-oriented arch bridge structure can greatly reduce the weight of the arch rib and is expected to break through the span of the arch bridge. In order to explore the construction technology of the new large-span arch bridge, the first UHPC box arch bridge in China was designed and built. Through the design, construction and experiment of a small span bridge over Yunnan Changbao Expressway, the key technologies of economical UHPC preparation and UHPC arch bridge construction are studied. The research results show that : the use of UHPC to construct the arch bridge can reduce the thickness of the section and greatly reduce the weight of the arch rib; UHPC made of machine-made sand can not only meet the requirements of bridge applications, but also reduce costs; the test results of the bridge show that the UHPC arch bridge has good mechanical performance. The bridge fully demonstrates the application potential of UHPC in arch bridges, and provides technical accumulation for the subsequent construction of large-span arch bridges.

Keywords: bridge engineering; box arch bridge; vehicle flyover; ultra high performance concrete (UHPC); machine-made sand

抗风缆加强型悬索桥合理成桥状态确定与力学性能研究

刘 铮 孙 斌
（同济大学 土木工程学院，上海，200092）

摘　要：本文将抗风缆加强型悬索桥划分为悬索结构和抗风缆结构，分别确定其合理成桥状态后再进行融合，从而建立了全桥的合理成桥状态。以某主跨为 246m 的双塔地锚式人行悬索桥为例，利用有限元分析确定了其平衡状态并分析了采用不同的抗风缆布置形式时结构的力学性能。分析结果表明：本文建立的合理成桥状态确定方法实施方便，且具有很高的精度；设置空间抗风缆能够有效限制主梁的侧向位移和主梁的竖向位移，从而改善主梁的受力状态。

关键词：抗风缆加强型悬索桥；合理成桥状态；力学性能；抗风缆倾角；抗风缆形式

Decision method on reasonable design state and research on mechanical performance of wind-resistant cable-reinforced suspension bridge

Liu Zheng　Sun Bin
(College of Civil Engineering, Tongji University, Shanghai 200092, China)

Abstract: In this paper, author divides the wind-resistant cable-reinforced suspension bridge into a suspension cable structure and a wind-resistant cable structure. The reasonable design state of the suspension bridge is determined separately before fusion, thereby establishing a reasonable design state of the full bridge. Taking a two-tower ground-anchored pedestrian suspension bridge with a main span of 246m as an example, the reasonable design state is determined by finite element analysis, and the mechanical properties of the structure with different wind-resistant cable arrangements are adopted and analyzed. The analysis results show that the method for determining the reasonable design state established in this paper is easy to implement and has high accuracy. The installation of the spatial wind-resistant cables can effectively limit the lateral displacement of the main girder and the vertical displacement of the main girder, thereby improving the performance of the structure.

Keywords: wind-resistant cable reinforced suspension bridge; reasonable design state; mechanical properties; wind-resistant cable inclination; wind-resistant cable form

超大断面浅埋隧道双侧壁导坑法及其优化设计

刘春舵　杨超超　刘忠凯　宁朝阳　赵长龙
（中海建筑有限公司，广东 深圳，510000）

摘　要：在超大断面浅埋隧道施工中，使用双侧壁导坑法能有效控制围岩变形，对保证施工安全具有十分重要的意义。以贵州黔南地区Ⅴ级围岩的师院隧道为工程背景，运用有限元软件 MIDAS/GTS 研究了超大断面浅埋隧道采用双侧壁导坑六步开挖工法进行施工时围岩变形沉降规律。根据模拟结果，进行了双侧壁导坑九步法的优化设计。研究表明：对于Ⅴ级围岩的超大断面浅埋隧道，采用双侧壁导坑九步法施工，能有效控制围岩变形；采用双侧壁导坑九步法与六步法在开挖过程中的围岩变形规律是相似的；在这类隧道施工过程中，左导坑下部、右导坑上部和核心土上部开挖导致围岩变形突变、沉降位移突增，为围岩稳定性的主要控制点。隧道监测表明：使用双侧壁导坑九步法施工，再配合辅助措施在减小围岩沉降变形上是可行的。研究结果为对Ⅴ级围岩的超大断面浅埋隧道施工设计提供了一定的参考借鉴。

关键词：浅埋隧道；超大断面；双侧壁导坑法；围岩变形；优化设计；数值模拟

Double side slope method and its optimization design in shallow and super-large section tunnel

Liu Chunduo　Yang Chaochao　Liu Zhongkai　Ning Chaoyang　Zhao Changlong
(China Shipping Construction Co., Ltd., Shenzhen, Guangdong 510000, China)

Abstract: In the construction process of shallow and super-large section tunnels, the use of the double side slope method can effectively control the deformation of surrounding rocks, which is of great significance for ensuring construction safety. With the engineering background of the Ⅴ level surrounding rock in Qiannan area of Guizhou Province, the finite element software MIDAS/GTS was used to study the deformation and displacement law of surrounding rock during the construction of a shallow and super-large section tunnel using the double side slope method. Based on the simulation results, the optimization design of the double side slope method by nine steps was carried out. The research shows that: for the shallow and super-large section tunnel with Ⅴ level surrounding rock, the use of the double side slope method by nine steps can effectively control the deformation of surrounding rock. The deformation law of surrounding rock is similar during the construction of the tunnel using the double side slope method by nine steps or six steps. During the construction of the type of the tunnels, the excavation of the lower part of the left drift, the upper

part of the right drift and the upper part of the core rock caused sudden deformation of the surrounding rock and sudden increase in settlement displacement, which are the main control points for the stability of the surrounding rock. The tunnel monitoring shows that the double side slope method by nine steps, combined with auxiliary measures, is feasible to reduce the deformation of surrounding rocks. The research results can provide some reference for the construction design of the shallow and super-large section tunnels in the V level surrounding rocks.
Keywords: shallow-buried tunnel; super-large section; double side slope method; surrounding rock deformation; optimization; numerical simulation

高瓦斯隧道压入式通风风带选型及安装位置优化研究

郑仕跃

（中海建筑有限公司，广东 深圳，518000）

摘　要：为优化高瓦斯隧道施工通风参数，采用仿真模拟与实践相结合的探究方式，探究影响瓦斯隧道稀释通风风流场中掌子面附近瓦斯浓度场分布的相关因素的影响规律，探讨合理的风带位置、风带口直径、不同隧道断面、不同出口风速，风带口流速等气流组织形式，对特定瓦斯涌量情况下的瓦斯浓度场分布情况进行仿真模拟，找出最佳参数，将掌子面的瓦斯浓度控制在一定范围之内。在实际施工通风中按照模拟得出的参数，不同施工工况，有效通风方式，确保瓦斯浓度在规范允许安全限制之内。

关键词：高瓦斯隧道；施工通风；通风方式；通风参数

Study on the selection and installation position optimization of press-in ventilation belt in high gas tunnel

Zheng Shiyue

(China Overseas Construction Co. , Ltd. , Shenzheng, Guangdong 518000, China)

Abstract: In order to optimize the construction ventilation parameters of high gas tunnels, a combination of simulation and practice is adopted to explore the influence law of related factors that affect the distribution of gas concentration field near the tunnel face in the dilution ventilation air flow field of gas tunnels, and to explore reasonable wind Airflow organization forms such as belt location, wind belt opening diameter, different tunnel sections, and different exit wind speeds, wind belt opening flow velocity, etc. , simulate the gas concentration field distribution under specific gas influx, find the best parameters, and The gas concentration of the tunnel face is controlled within a certain range. In actual construction ventilation, according to the simulated parameters, different construction conditions, and effective ventilation methods, ensure that the gas concentration is within the safety limits allowed by the specification.

Keywords: high gas tunnel; construction ventilation; ventilation method; ventilation parameters

大纵坡曲线条件下钢箱梁桥结构受力分析

刘　瑶　邹德强　李伟东

（中国建筑第五工程局有限公司，湖南 长沙，410004）

摘　要：为了研究大纵坡曲线条件下连续刚构桥的受力特征，以国内某（51.6+51.6）m 跨线钢箱梁桥为例，建立了考虑桩基-桥墩-钢箱梁一体化有限元模型，系统分析结构在恒载、活载、温度、不均匀沉降等荷载作用下的受力性能。结果表明：对于曲线桥梁，单梁模型仅为平均值，无法反映主梁受力的最大值，建议采用梁格法更安全；恒载、活载作用下，墩顶位置中间主梁受力最大，其余位置外侧主梁应力最大；0＃墩内侧支座最大反力比外侧小 72.4%；整体温度、不均匀沉降对结构的应力影响较小；温度梯度作用下，主梁最大应力为 16.5MPa；自墩底 0～10m 范围内中墩变形随着高度呈线性变化，10～18m 范围内中墩变形随高度呈抛物线变化；墩柱最大轴力沿竖向线性增加，墩柱弯矩、剪力沿竖向为恒值；桥梁第一阶振型为正对称扭转，对应频率为 2.3Hz。

关键词：工业化；双结合；钢板组合梁；受力分析

Structural stress analysis of steel box girder bridge with large longitudinal slope curve

Liu Yao　Zou Deqiang　Li Weidong

(China Construction Fifth Engineering Division Co., Ltd., Changsha, Hunan 410004, China)

Abstract: To study the mechanical characteristics of continuous rigid frame bridge under the condition of large longitudinal slope curve, a steel box girder bridge with the span of (51.6+51.6) m is taken as an example. The finite model is established by pile foundation, pier, steel box girder. The paper systematically analyzed the mechanical performance under dead load, live load, temperature, uneven settlement and other loads. The conclusions are obtained: For curved bridges, the single beam model is only the average value, which can not reflect the maximum stress of the main beam, so the grillage method is recommended to be more safe. Under the action of dead load and live load, the stress of the main beam in the middle of the pier top is the largest, and that of the main beam outside the other positions is the largest. The maximum reaction force of the inner support of 0 # pier is 72.4% smaller than that of the outer support. The overall temperature and uneven settlement have little influence on the stress of the structure. Under the action of temperature gradient, the maximum stress of main girder is 16.5MPa. From the bottom of the

pier, the deformation of the middle pier changes linearly with the height in the range of 0-10m, and parabola in the range of 10-18m. The maximum axial force of the pier increases linearly along the vertical direction, and the bending moment and shear force of the pier are constant along the vertical direction. The first mode shape of the bridge is positive symmetric torsion, and the corresponding frequency is 2.3Hz.
Keywords: industrialization; double-combination; steel plate composite beam; analysis

空间索面斜拉桥索导管精确定位技术

张　欢[1,2]　李　璋[1]　周　帅[1]　雷　军[1]　何昌杰[1]　方　聪[1]
（1. 中国建筑第五工程局有限公司，湖南 长沙，410004；
2. 中建五局总承包公司，湖南 长沙，410004）

摘　要：为提高空间索面斜拉桥索导管定位的精度和施工的便利性，对索导管定位技术进行研究。将索导管两端的斜切口使用钢板封住，并将索导管的空间中轴线与斜切口的交点印刻在钢板上，实现虚拟控制点实体化，进而实现三维坐标直接定位。将该技术应用于某斜拉桥，应用结果表明该定位技术科学合理，定位速度快，定位精度满足±5mm 的设计要求，定位方法易于推广应用。
关键词：空间索面；斜拉桥；索导管；精确定位；控制点

Accurate positioning technique of cable conduits in cable-stayed bridges with space cables

Zhang Huan[1,2]　*Li Zhang*[1]　*Zhou Shuai*[1]　*Lei Jun*[1]　*He Changjie*[1]　*Fang Cong*[1]
(1. China Construction Fifth Engineering Bureau Co. Ltd., Changsha, Hunan, 410004, China; 2. General Contracting Company of China Construction Fifth Engineering Bureau, Changsha, Hunan 410004, China)

Abstract: In order to improve the positioning accuracy of cable conduits and the convenience of construction of cable-stayed bridges with space cables, the accurate positioning technique for cable conduits of cable-stayed bridges with space cables is studied. The oblique incisions at both ends of the cable conduits are sealed with steel plate, and the intersection of the spatial central axis of the cable conduits with the oblique incision is engraved on the steel plate to transform the virtual control points to to actual ones, and then position the three-dimensional coordinates directly. The technique has been used to a cable stayed bridge. The application results show that the positioning technique is scientific and reasonable, which can make the cable conduits position quickly, and the positioning accuracy is within 5mm. The technique is easy to be popularized.
Keywords: space cables; cable-stayed bridges; cable conduits; accurate positioning; control point

LRB在高震区曲线高架桥中的适用性研究

李　璋[1]　周　帅[1]　李　凯[1]　谭芝文[2]　罗桂军[3]　杨　坚[1]
（1. 中国建筑第五工程局有限公司，湖南 长沙，410004；
2. 中建隧道建设有限公司，重庆，401320；
3. 中建五局土木工程有限公司，湖南 长沙，410004）

摘　要：为了研究铅芯隔震橡胶支座（LRB）在曲线钢箱梁和混凝土梁过渡时的适用性，使用有限元软件Midas Civil建立三维桥梁模型，包括盆式和铅芯两种橡胶支座，对两种模型的静力、动力性能进行深入对比研究。分析结果表明，间接荷载作用下采用LRB时墩底横向弯矩较小；而在直接荷载作用下，采用LRB时，单个桥墩内力较大。LRB的设置使结构的位移更加敏感，当两联之间结构刚度、重量等参数的差异较大时，应重点关注两联之间的位移差。LRB的采用，可使所研究的曲线高架桥纵桥向减震率介于7%～43%，横桥向减震率介于56%～76%，具有良好的减震效果。

关键词：LRB；曲线高架桥；高震区；非线性时程分析；应用性

Applicability study of LRB in Curved Viaduct in high earthquake area

Li Zhang[1]　*Zhou Shuai*[1]　*Li Kai*[1]　*Tan Zhiwen*[2]　*Luo Guijun*[3]　*Yang jian*[1]
(1. China Construction Fifth Engineering Bureau Co. Ltd., Changsha Hunan 410004, China; 2. China Construction Tunnel Co. Ltd., Chongqing 401320, China, 3. Civil Engineering Co. Ltd. of China Construction Fifth Engineering Bureau, Changsha, Hunan 410004, China)

Abstract: To investigate the applicability of Lead-rubber bearing (LRB) in the transition between curved steel box girder and concrete beam, three-dimensional models of bridge structures with basin type rubber bearings and Lead-rubber bearings were established by using the finite element software MIDAS civil, and their static and dynamic performances were studied. The results show that the transverse bending moment at the bottom of pier is smaller under indirect load when LRB is used. Otherwise, the internal force of single pier is larger when the Lead-rubber bearing is used under the direct loads. The setting of the Lead-rubber bearing makes the displacement of the structure more sensitive. When the stiffness, weight or other parameters of the structural are greatly different between the two couplers, we should pay attention to the displacement difference between the two couplers. The longitudinal and transverse vibration reduction rates of the Curved Viaduct studied are between 7%-43% and 56%-76%. Lead-rubber bearing has excellent aseismic performance.

Keywords: LRB; curved viaduct; high earthquake area; time-history analysis; applicability

大跨径梁拱组合刚构桥下弦拱梁挂篮选型及施工应用

秦宗琛　李亚勇　张　斌　王　蓬　张　锐
（中建隧道建设有限公司，重庆，230009）

摘　要： 梁拱组合混凝土刚构桥是一种由梁拱融合而成的新型桥梁结构形式，主梁为上梁、下拱组成三角区结构，其下弦拱梁线型变化大，斜拱角度大。施工时会遇到拱梁“大坡度、变弧线、挂篮斜爬”等施工难题。以礼嘉嘉陵江大桥工程下弦拱梁为例开展挂篮优化选型研究，形成下承式倒三角挂篮悬浇施工技术。此技术成功应用表明：下承式挂篮爬坡能力强，拱梁线型质量高、施工效率高。该工艺在大坡度混凝土拱梁悬臂施工取得多方面成果，为此类型桥梁施工提供成熟的施工技术。

关键词： 梁拱组合刚构桥；挂篮悬臂浇筑；下承式挂篮；下弦拱梁

The type selection and construction application of hanging basket for lower chord arch beam of large span beam-arch composite rigid frame bridge

Qin Zongchen　Li Yayong　Zhang Bin　Wang Peng　Zhang Rui
(China Construction Tunnel Corp., Ltd, Chongqing 230009, China)

Abstract: beam-arch composite concrete rigid frame bridge is a new type of bridge structure formed by the combination of beam and arch. The main girder is a triangular structure composed of upper girder and lower arch. The lower chord arch beam has a large change in line type and large angle of inclined arch. During construction, construction problems such as " steep slope, arc-changing curve and ramp climbing of cradles" will be encountered. Taking the lower chord arch beam of Lijia-Jialingjiang Bridge as an example, the study on optimal selection of cradles is carried out to form the suspension casting construction technology of undercarriage inverted triangular cradles. Successful application of this technology shows that the undercarriage has strong climbing ability, high linear quality of arch beam and high construction efficiency. This technology has achieved many results in cantilever construction of concrete arch beam with large slope, and provides mature construction technology for this type of bridge construction.

Keywords: beam-arch composite rigid frame bridge; hanger cantilever casting; lower-carrying cradle; lower chord arch

城市山岭隧道穿越断层破碎带的关键施工技术研究

陈宇波　李佳文　詹树高　余　浪
（中建隧道建设有限公司，重庆，401320）

摘　要： 为了研究城市山岭隧道穿越断层破碎带的关键施工技术，以重庆市快速路二横线土主隧道工程为例，参考类似工程案例，提出采用帷幕注浆预加固方案，分析了不同注浆参数下的注浆效果。针对土主隧道的水文地质情况，选择合适的注浆参数，取得了良好的注浆固结效果，安全穿越了断层破碎带，为今后类似工程提供科学依据。
关键词： 城市山岭隧道；断层破碎带；全断面帷幕注浆；注浆参数；关键施工技术

Research on key construction technology of urban mountain tunnel crossing fault fracture zone

Chen Yubo　Li Jiawen　Zhan Shugao　Yu Lang
(China Construction Tunnel Construction Co., Ltd., Chongqing 401320, China)

Abstract: To research the key construction technology of urban mountain tunnel crossing fault fracture zone, taking the Tu-zhu Mountain tunnel project of Chongqing suburban railway as an example, referring to similar engineering cases, the curtain grouting pre-reinforcement scheme was proposed to analyze the grouting effect under different grouting parameters. In view of the hydrogeological conditions of the Tu-zhu Mountain Tunnel, the appropriate grouting parameters were selected to safely cross the fault fracture zone, providing a scientific basis for similar projects in the future.
Keywords: urban mountain tunnel; fault fracture zone; full-section curtain grouting; grouting parameters; key construction technology

浅埋暗挖隧道穿越复杂建筑物安全风险模糊评价

梁　军　尹　辉　张浩铌　钟菊焱　陈　晨
（中建隧道建设有限公司，重庆，401320）

摘　要： 城市隧道开挖引起的周围环境安全问题，尤其是周边建筑物安全问题越来越受到人们的广泛关注。影响隧道临近既有建筑物风险因素比较多，为了能够对其风险程度作出较为客观的评价，针对当前隧道施工导致建筑损伤风险因素仍存在许多不确定性与模糊性的特点。本文提出模糊层次分析方法评估隧道穿越建筑群的施工风险概率估计及邻近区域某一特定建筑的损伤风险。通过工程实例验证，提出的方法能有效准确地对建筑物风险的不同程度进行评价，初步实现了评价从定性、经验性向科学化和半定量化的转变，研究成果可在今后类似工程中推广应用，对工程施工具有一定的指导和借鉴意义。
关键词： 隧道；浅埋暗挖；地表建筑；风险；模糊综合评价

Fuzzy evaluation on safety risk of shallow buried tunnel passing through complex buildings

Liang Jun　Yin Hui　Zhang Haoni　Zhong Juyan　Chen Chen
(China Construction Tunnel Construction Co., Ltd., Chongqing 401320, China)

Abstract: The surrounding environmental safety problems caused by the excavation of urban tunnels, especially the safety problems of surrounding buildings, have been paid more and more attention, there are many risk factors that affect the existing buildings near the tunnel, in order to be able to make a more objective evaluation of its risk level, in view of the current tunnel construction leading to building damage risk factors still have many uncertainties and fuzzy characteristics, this paper proposes a fuzzy analytic hierarchy process method to evaluate the construction risk probability of tunnel passing through buildings and the damage risk of a particular building in the adjacent area. Through the verification of engineering examples, the proposed method can effectively and accurately evaluate different degrees of building risk, and preliminarily realize the transformation of evaluation from qualitative and empirical to scientific and semi quantitative. The research results can be popularized and applied in similar projects in the future, which has certain guiding and reference significance for engineering construction.
Keywords: tunnel; shallow buried excavation; surface building; risk; fuzzy comprehensive evaluation

正交异性钢桥面板横隔板处弧形开孔对比优化研究

严德华[1]　杨　羿[2]　刘　朵[2]　张建东[1,2]

(1. 南京工业大学 土木工程学院，江苏 南京，211800；2 苏交科集团股份有限公司在役长大桥梁安全与健康国家重点实验室，江苏 南京，211112)

摘　要：横隔板处弧形开孔是影响正交异性钢桥面板疲劳性能重要构造参数之一。本文通过 ABAQUS 建立精细化有限元模型，模拟横隔板弧形开口处真实的应力状态，确定不同孔型最不利加载位置，对比了不同弧形开孔形式对主应力的影响。结果表明：弧形开孔区域 U 肋和横隔板焊缝端部以及开孔自由边为疲劳易损部位，与实际桥梁疲劳裂纹一致；U 肋和横隔板焊缝端和弧形开孔自由边处最不利横桥向荷位相同为距跨中横隔板 150mm 处，最不利纵桥向荷位分别距跨中横隔板 600mm 和 200mm；六种孔型中，综合比较第一主应力和第三主应力值，方案六最优；弧形开孔面积或开孔高度的增大会减小 U 肋和横隔板焊缝端部的第一主应力值，开孔半径的增大会降低开孔自由边处的第三主应力值。

关键词：正交异性钢桥面板；疲劳裂纹；弧形开孔；主应力

Comparative optimization of arc opening at diaphragm of orthotropic steel bridge deck

Yan Dehua[1]　*Yang Yi*[2]　*Liu Duo*[2]　*Zhang Jiandong*[1,2]

(1. Nanjing Tech University, Nanjing, Jiangsu 211800, China;

2 . The State Key Laboratory on Safety and Health of In-service Long-span Bridge, JSTI Group, Nanjing, Jiangsu 211112, China)

Abstract: Arc opening at diaphragm is one of the important structural parameters that affect the fatigue performance of orthotropic steel bridge deck. In this paper, a refined finite element model is established by ABAQUS to simulate the real stress state at the arc opening of diaphragm, determine the most unfavorable loading position of different hole types, and compare the influence of different arc opening forms on the principal stress. The results show that the weld ends of U-rib and diaphragm in the arc-shaped opening area and the free edge of the opening are vulnerable parts, which are consistent with the actual bridge fatigue cracks; The most disadvantageous transverse bridge load position at the weld end of U-rib and diaphragm and the free edge of arc opening is the same, which is 150 mm away from the diaphragm in the middle of span, and the most disadvantageous longitudinal bridge load position is 600 mm and 200 mm away from the diaphragm in the middle of span respectively; Among the six pass patterns, scheme 6 is the best; The first principal stress at the end of U-rib and diaphragm weld decreases with the increase of arc opening area or opening height, and the third principal stress at the free edge of opening decreases with the increase of opening radius.

Keywords: orthotropic steel deck; fatigue crack; arc opening; principal stress

过江小净距隧道暗挖陆域段开挖方法及安全步距研究

罗桂军[1]　傅鹤林[2]　郭弘宇[1]　雷润杰[1]

（1. 中建五局土木工程有限公司，湖南 长沙，410004；2. 中南大学，湖南 长沙，410004）

摘　要：为解决小净距大断面软岩隧道开挖过程中，相邻两峒室产生相互扰动而存在的施工安全问题，依托蓉江四路过江隧道暗挖陆域段工程，在前期围岩注浆预加固施工基础上，运用FLAC3D数值模拟软件，计算三台阶法、CD法和CRD法3种施工方法在不同步距（10m、20m、30m、40m）下的施工力学行为。计算结果表明：后行隧道开挖对先行隧道的位移变形存在有利影响。两峒室之间的开挖步距越大，后行隧道对先行隧道的影响越小。开挖台阶长度相同时，开挖界面越小，围岩所产生的位移越小。步距相同的情况下，在初期支护变形与围岩位移控制效果上，CRD法>CD法>三台阶法。从各工序转换及施工功效看，采用CD法开挖步距更为合理，效率更高。

关键词：过江小净距隧道；数值模拟；开挖方法；安全步距

Study on excavation method and safe distance of small clear distance tunnel in land area across river

Luo Guijun[1]　*Fu Helin*[2]　*Guo Hongyu*[1]　*Lei Runjie*[1]

（1. China Construction No. 5 Engineering Bureau, Changsha, Hunan 410031, China;
2. Central South University, Changsha, Hunan 410075, China）

Abstract: In order to solve the safety problem of the two adjacent caverns caused by mutual disturbance during the excavation of soft rock tunnel with small clear distance and large section, based on the underground excavation land section project of Rongjiang No. 4 Road Tunnel, the construction mechanics behavior of three construction methods (three-step method, CD method and CRD method) under the non-synchronous distance (10m, 20m, 30m and 40m) is calculated by using FLAC3D numerical simulation software on the basis of the pre-reinforcement construction of surrounding rock grouting in the early stage. The results show that the excavation of the back tunnel has a beneficial effect on the displacement of the first tunnel. The larger the excavation distance between two chambers, the less the influence of the back tunnel on the first tunnel. When the length of excavation steps is the same, the smaller the excavation interface, the smaller the displacement of surrounding rock. With the same step length, CRD method>CD method>three-step method is used to control the initial deformation of support and the displacement of surrounding rock . CRD method has the best effect on controlling ground subsidence and vault subsidence.

Keywords: small clear distance tunnel across the river; numerical simulation; excavation method; safety interval

专题七　地下空间高效开发与利用

谐波激励下软土场地框架式地铁车站动力响应试验研究

张志明[1,2]　袁　勇[3]

（1. 贵州大学 土木工程学院，贵州 贵阳，550025；
2. 贵州省岩土力学与工程安全重点实验室，贵州 贵阳，550025；
3. 同济大学 土木工程防灾国家重点实验室，上海，200092）

摘　要：为研究谐波激励下软土场地框架式地铁车站的动力响应，开展了 1/30 比例尺的土-框架式地铁车站振动台模型试验。采用人工模型土、镀锌钢丝和微粒混凝土分别模拟原型软土场地、钢筋和混凝土。通过改变振动台输入地震动频率，探究其对框架式车站及周围场地动力响应的影响规律，分析土-结构系统的加速度及其传递函数、卓越频率、结构动应变、侧墙动土正应力等。试验结果显示：（1）土-框架式车站模型的一阶频率与自由场地的一阶频率一致，印证了地铁车站的地震响应由场地控制；（2）侧墙动土正应力的频谱曲线反映场地加速度响应传递函数曲线的特征，最大动土正应力发生在侧墙最底端；（3）车站结构动应变频谱曲线反映场地加速度响应传递函数曲线的特征，谐波激励下车站站台层柱顶动拉应变最大，建议增强站台层柱顶的抗震性能。

关键词：框架式地铁车站；动力响应；振动台试验；谐波激励

Experimental study ondynamic responses of frame subway station in soft soil under harmonic excitations

Zhang Zhiming[1,2]　*Yuan Yong*[3]

（1. College of Civil Engineering，Guizhou University，Guiyang，Guizhou 550025，China；
2. Guizhou Provincial Key Laboratory of Rock and Soil Mechanics and Engineering Safety，Guizhou University，Guiyang ，Guizhou 550025，China；3. State Key Laboratory of Disaster Reduction in Civil Engineering，Tongji University，Shanghai 200092，China）

Abstract：To investigate the dynamic responses of frame subway station in soft soil under harmonic excitations，shaking table tests were performed on a 1/30 scaled soil-frame subway station model. The artificial soil，galvanized steel wire and micro-concrete were used to model the ground，steel rebar and concrete of the prototype，respectively. By changing the frequencies of input motions，influence mechanism of the frequencies on the dynamic responses of both frame subway station and surrounding ground was explored. Experimental data of soil-structure system were analyzed，including the acceleration，transfer function，predominant frequency，structural dynamic strain，and dynamic soil normal stress a-

long the sidewall. The test results revealed that: (1) The first predominant frequency of soil-frame subway station model is consistent with that of free-field model, which proves that the seismic responses of subway station are controlled by the surrounding ground; (2) The spectral characteristics of dynamic soil normal stress reflect the transfer function of the ground acceleration and the maximum stress occurs at the bottom of sidewall; (3) The spectral characteristics of dynamic structural strain reflect the transfer function of the ground acceleration. Under harmonic excitations, the maximum dynamic structural strain occurs on the top of the column belong to the platform floor. It is advised to strengthen the seismic performance of the column top.
Keywords: frame subway station; dynamic response; shaking table test; harmonic excitation

松散地层条件下的盾构带压开仓技术研究

黄立辉　贾建伟　周　伟　孟令冲
（中建六局轨道公司 安徽 合肥，300450）

摘　要： 随着城市轨道交通的发展，盾构机作为一种先进、安全、可靠的施工技术而被广泛应用。在盾构掘进施工过程中，如盾构机刀具磨损严重开仓换刀便成了唯一的选择，但在开仓换刀过程中，确保掌子面的稳定是施工关键技术之一。盾构带压开仓作业作为一种重要的开仓工艺，其前提条件是地层或经过处理的地层具备一定的自稳能力及具备一定的气密性，能够通过在气压的辅助作用下较长时间地为开仓人员进入土仓提供一个稳定隔水作业环境，在气密性较好的黏土类地层中应用广泛。但在松散、孔裂隙发育及气密性差的回填土、孔裂隙发育的黏性土等地层鲜有成功案例。本文结合厦门轨道 4 号线工程在松散、孔裂隙发育及气密性差的特殊地质环境中在气压无法较长时间保证稳定的条件下，通过在开仓过中动态向土仓内充气使仓内气压与水压保持一个平衡状态，从而为盾构机土仓内提供一个稳定的隔水环境从而进行带压开仓换刀，为类似工程案例积累施工经验。
关键词： 带压开仓；松散地层；孔裂隙发育；盾构

Study on the technology of shield tunnel opening under the condition of loose stratum

Huang Lihui　Jia Jianwei　Zhou Wei　Meng Lingchong
(China Construction Sixth Engineering Bureau Railway Transportation Corporation, Hefei, Anhui 300450, China)

Abstract: With the development of urban rail transit, shield tunneling machine is widely used as an advanced, safe and reliable construction technology. In the construction process of shield tunneling, if the knife of shield machine wears severely, opening and changing the knife of shield machine has become the only option. However, in the process of opening and changing the knife of shield machine, ensuring the stability of the tunnel face is one of the key construction techniques. Shield opening under pressure is an important construction technology. The prerequisite is that the stratum or the treated stratum has a certain degree of self-stability and a certain degree of air tightness. It can be compared with the help of air pressure. It provides a stable water-proof working environment for the openers to enter the soil bin for a long time, and is widely used in clay-like formations with good airtightness. However, there are few successful cases in the backfill soil with loose, well-developed pores and cracks and poor air tightness, and cohesive soil with well-developed

pores and cracks. This article combines the Xiamen Rail Line 4 project in the special geological environment with loose, pores and fissures and poor air tightness. Under the condition that the air pressure cannot be stabilized for a long time, the warehouse is dynamically filled with air during the opening of the warehouse. The air pressure and water pressure maintain a balanced state, so as to provide a stable water-proof environment in the soil bin of the shield machine to open the bin under pressure to change the knife, and accumulate construction experience for similar projects.
Keywords: shield machine open warehouse under pressure; loose formation; pore and fissure development; shield

上海软土深层地下工程试验基地建设程序及投融资分析

蔡国栋[1]　刘千诚[2]　白　云[3]

（1. 上海勘察设计研究院（集团）有限公司，上海，200093；

2. 上海国际工程咨询有限公司，上海，200093；

3. 上海瓴云土木工程咨询有限公司，上海，200333）

摘　要： 为更好地解决上海市建筑用地不足、交通拥堵以及环境恶化等突出问题，深层地下空间（深度大于40m）开发成为优化城市空间结构、提高城市空间资源利用效率的重要手段。深层地下空间开发需要新装备、新材料、新工艺的支撑，而上述新型技术需要经过工程验证才能在实际工程中进行应用，因此，建设上海市软土深层地下工程试验基地，为新装备、新材料、新工艺提供工程验证场所显得尤为重要。本文从试验基地建设程序、投融资模式、试验基地后续利用进行研究分析，为试验地基顺利建设与运营奠定基础。

关键词： 软土；试验基地；建设程序；投融资模式

Construction procedure, investment and financing analysis of Shanghai soft soil deep underground engineering test site

Cai Guodong[1]　*Liu Qiancheng*[2]　*Bai Yun*[3]

（1. SGIDI Engineering Construction（Group）Co.，Ltd.，Shanghai 200093，China；

2. Shanghai International Engineering Consulting Co.，Ltd.，Shanghai 200093，China；

3. BY Civil Engineering Consulting Co.，Ltd.，Shanghai 200333，China）

Abstract：In order to solve the shortage of building land，traffic congestion，environmental degradation and other prominent problems in Shanghai，the development of deep underground space（The depth is greater than 40 meters）has become an important way to optimize the urban spatial structure and improve the utilization efficiency of urban spatial resources. Deep underground space development need new equipment，new material and new process，and those new technologies require engineering verification in the practical engineering application，therefore，it is important to construct deep underground engineering test site of soft soil in Shanghai which providing the site to evaluate new equipment，new material and new process. In this paper，the construction procedure，investment and financing mode and subsequent utilization of the test site are studied and analyzed in order to lay the foundation for the construction and operation of the test site.

Keywords：soft soil；test site；construction procedure；investment and financing mode

“以道定形”的城市地下公共服务空间规划开发利用构想

李　庆
（中铁第四勘察设计院集团有限公司，湖北 武汉，430000）

摘　要：随着城市逐步往地下发展，城市地下公共服务空间的规划模式成为城市规划的重点。本文通过城市地下公共服务空间规划的设计实践，探索城市的发展历史，借鉴河道、马车道、轨道等在不同时期扮演重要角色的模式。分析城市地下公共服务空间的关键要素，提出以“轨道、步道、绿道、车道”为城市地下空间规划之“道”，在“道”的基础上推演出城市地下人行道、市政管廊、商业服务、地下环路、物流管道等公共服务空间的规划之“形”。针对困扰建设方的“合理规划布局城市地下公共服务空间”问题提出一种规划模式。

关键词：“以道定形”；城市地下公共服务空间；地下空间规划；地下空间形式；地下空间利用模式

Conception of urban underground public service space planning, development and utilization

Li Qing
(China Railway Fourth Survey and Design Institute Group Co. , Ltd. , Wuhan, HuBei 430000, China)

Abstract: With the gradual development of the city underground, the planning mode of urban underground public service space has become the focus of urban planning. Through the design practice of urban underground public service space planning, this paper explores the development history of the city, and draws lessons from the model of river, carriage way, rail and so on, which play an important role in different periods. This paper analyzes the key elements of urban underground public service space, and puts forward that "track, footpath, greenway and driveway"should be taken as the " road" of urban underground space planning. On the basis of "road", the "shape" of urban underground sidewalk, municipal pipe gallery, commercial service, underground Ring Road, logistics pipe and other public service space planning should be promoted. In view of the problem of "reasonable planning and layout of urban underground public service space", a planning mode is proposed.

Keywords:"Shape by Tao"; urban underground public service space; underground space planning; underground space form; utilization mode of underground space emi-rigid connection

广州地铁纪念堂暗挖车站分层与深层土体沉降监测分析

徐顺明[1]　陈建党[1]　陈巨武[1]　彭丕洪[2]
（1 广州地铁集团有限公司，广东 广州，510380
2. 广州地铁设计研究院股份有限公司，广东 广州，510330）

摘　要： 广州地铁拱盖法暗挖车站较多，由于暗挖车站顶部存在砂层、淤泥质黏土、粉质黏土等软弱地层，其普遍具有高含水量性、高压缩性、低透水性和低强度等特点，施工阶段易造成较大变形，将会引起隧道坍塌和周边建构筑物设施破坏，造成严重的生命与财产损失，继而造成重大的社会影响，文章以纪念堂站为监测对象，增设分层、深层土体监测项目对软弱地层暗挖车站的沉降规律进行分析和总结得出有益结论，为暗挖车站施工安全提供依据。

关键词： 暗挖车站；拱盖法施工；深层土体；监测 ；分析

Monitoring and analysis of the layered and deep soil settlement of the underground excavation station in Guangzhou Metro Memorial Hall

Xu Shunming[1]　*Chen Jiandang*[1]　*Chen Juwu*[1]　*Peng Pihong*[2]
(1. Guangzhou Metro Group Co., Ltd., Guangzhou, Guangdong 510330, China;
2. Guangzhou Metro Design & Research Institute Co., Ltd., Guangzhou,
Guangdong 510330, China)

Abstract: There are many underground excavation stations in Guangzhou Metro arch cover method. Because of the soft stratum such as sand layer, silt clay and silty clay at the top of the underground excavation station, it has the characteristics of high water content, high pressure shrinkage, low permeability and low strength. It is easy to cause large deformation during construction, which will cause tunnel collapse and damage of surrounding buildings and facilities, and cause serious life and property losses, Then it has a great social impact. Taking Memorial Hall Station as the monitoring object, the article adds layered and deep soil monitoring items to analyze and summarize the settlement law of the weak ground underground excavation station, and draw useful conclusions, which provides the basis for the construction safety of the underground excavation station.

Keywords: underground excavation station; arch cover construction; deep soil; monitoring and analysis

复杂环境下埋地管道非大开挖穿越工程施工技术

汤 毅 王建勃 曹晓程 施 强 吴增强
（上海市安装工程集团有限公司，上海，200080）

摘 要：非大开挖施工作业模式可减少对环境、工期、周边管线等影响，是未来埋地管线施工技术的发展方向。本文对某埋地油气管线非大开挖施工技术做了介绍，对其关键工序及施工重点做了说明，希望对类似工程起到借鉴作用。
关键词：非大开挖；管线；钻土；泥浆；回拖

Construction technology of buried pipeline crossing by non-excavation in complex environment

Tang Yi　Wang Jianbo　Cao Xiaocheng　Shi Qiang　Wu Zengqiang
(Shanghai Installation Engineering Co., Ltd., Shanghai 200080, China)

Abstract: The non-large excavation construction mode can reduce the impact on the environment, construction period and surrounding pipelines, etc., and is the development direction of buried pipeline construction technology in the future. This paper introduces the non-large excavation construction technology of a buried oil and gas pipeline, explains its key working procedure and key construction points, and hopes to provide reference for similar projects.
Keywords: non-large excavation; pipeline; drill soil; mud; drag back

隐伏溶洞等不良地质综合探测施工技术研究与应用

方 涛[1] 汪会来[1] 曹红军[1] 李朝辉[1] 张宏博[2]

（1. 中建五局第三建设有限公司，云南 大理，671000；
2. 中建铁路投资建设集团有限公司，湖北 武汉，430000）

摘 要：本文在综合分析钻孔取芯、压水试验、高密度电测、面波检测探测技术原理等内容的基础上，根据四种探测施工技术于铁路工程中不良地质的实际应用，并结合检测单位给出的一致性结构面探测结果报告，确定了岩溶注浆处理结果的可行性，以此提出了一种四种探测技术相结合检测不良地质的施工方法。该方法于实际工程的成功运用，定位定量性地解决隐伏溶洞等类似不良地质问题，确保了岩溶注浆的可靠性、连续性、整体性，进而证实了本文提出的隐伏溶洞等不良地质综合探测施工技术的可行性，可以为类似工程问题提供参考依据。

关键词：钻孔取芯；压水试验；高密度电测；面波检测

Research and application of construction technology for comprehensive exploration of hidden karst cave and other unfavorable geology

Fang Tao[1] *Wang Huilai*[1] *Cao Hongjun*[1] *Li Zhaohui*[1] *Zhang Hongbo*[2]

（1. 3rd Construction Co.，Ltd of China Construction 5th Engineering Bureau，Changsha，Hunan 410007，China；2. China State Construction Railway Investment and Engineering Group Co.，Ltd.，Wuhan，Hubei 430000，China）

Abstract：Based on the comprehensive analysis of the principles of core drilling，water pressure test，high-density electrical measurement and surface wave detection，this paper determines the feasibility of karst grouting treatment results according to the practical application of four detection construction technologies in adverse geological conditions of railway engineering，combined with the detection results report of consistent structural plane given by the inspection unit，and puts forward a kind of four kinds of exploration The construction method of detecting bad geology by combining surveying technology. The successful application of this method in practical engineering can solve the similar geological problems such as hidden karst cave quantitatively，ensure the reliability，continuity and in-

tegrity of karst grouting, and further confirm the feasibility of the construction technology of comprehensive exploration of hidden karst cave and other adverse geological conditions proposed in this paper, which can provide reference for similar engineering problems.
Keywords: core drilling; water pressure test; high density electrical measurement; surface wave detection

大型换乘地铁车站超深永临结合钢管柱水下定位施工工法

郭　鹏[1]　于镦钧[2]

（1. 甘肃铁科建设工程咨询有限公司；2. 中交隧道局有限公司盾构分公司）

摘　要： 随着国内的城市地铁飞速发展，在市区繁华路段施工时，为减小对周围环境的影响，缓解市政交通压力，多采用全盖挖法或局部盖挖法施工。目前盖挖法多采用临时钢构柱、临时钢管柱或临时钢筋混凝土柱支撑盖挖顶板，临时立柱较多会造成盖挖区域下方土方开挖困难，后期拆除及修补成本较大，而永临结合钢管柱钢管内部填充 C50 微膨胀混凝土，具有刚度大、强度高等特点。基坑开挖时能支撑盖挖顶板、上方道路及交通荷载，车站结构施工时能兼做车站结构柱，大大节约了工程造价，减少了施工工序，节约狭窄的盖挖施工空间，有利于加快施工进度，适用绝大部分盖挖地铁车站施工。因此，近几年以来盖挖法竖向支撑系统采用永临结合钢管柱是一个发展趋势。

关键词： 换乘地铁车站；超深永临结合钢管柱；水下定位

Construction method of super deep permanent temporary connection and underwater positioning of steel pipe column in large transfer subway station

Guo Peng[1]　*Yu Dunjun*[2]

(1. Gansu Tieke Construction Engineering Consulting Co. , Ltd. ;
2. CCCC Tunnel Bureau Co. , Ltd. , Shield machine branch)

Abstract: With the rapid development of urban subway in China, in order to reduce the impact on the surrounding environment and alleviate the pressure of municipal traffic, the full-cover excavation method or partial cover excavation method is often used. At present, temporary steel structure column, Temporary Steel Pipe Column or temporary reinforced concrete column are often used in cover excavation method. More temporary columns will cause difficulties in Earth excavation under cover excavation area, and the later removal and repair cost is higher, while the C50 micro-expansion concrete filled in the steel tube of permanent and temporary steel tube column has the characteristics of high rigidity and high strength, it can support the roof, the road above and the traffic load when the foundation pit is excavated, and it can also be used as the station structural column during the construction of the station structure, it has greatly saved the cost of the project, reduced the construction procedure, saved the narrow construction space for cover and excavation,

helped to speed up the construction progress, and is suitable for the construction of the vast majority of cover and excavation subway stations. Therefore, in recent years, it is a developing trend to adopt permanent-temporary steel pipe column in the vertical supporting system of cover-excavation method.
Keywords: transfer to metro station; ultra-deep permanent and temporary joint steel pipe column; underwater positioning

地面铁路割裂背景下城市老城区对外地下步行系统规划及评价研究——以青岛市国际邮轮母港启动区及周边老城区为例

陆春方[1]　栾勇鹏[2]　董蕴豪[1]　彭芳乐[1]
（1. 同济大学 地下空间研究中心，上海，200092；
2. 青岛市人防建筑设计研究院有限公司，山东 青岛，266061）

摘　要：青岛市国际邮轮母港启动区与周边老城区受到铁路割裂影响，地面步行交通受到阻断，对老城区对外步行系统更新改造提出了新的要求。本文将广泛应用于城市规划和交通网络分析的空间句法作为步行交通系统评价手段，采用标准化角度整合度（NAIN）和标准化角度穿行度（NACH）作为分析步行空间集聚性和穿行性的量化指标，对邮轮母港片区与老城区之间的铁路割裂现状进行分析。依托既有地下工程、规划区域地下步行流线和空间句法计算得到的现状空间差异，提出了老城区对外地下步行交通系统规划方案。最后对不同方案下的过境交通和到发交通吸引力进行了评估。本文结合案例分析希望为老城区在铁路割裂下的地下步道更新改造提供借鉴。此外通过空间句法的应用为国内老城区步行系统的规划和评价提供示范。

关键词：地下步行交通；铁路割裂；老城区；空间句法

Planning and evaluation of external underground pedestrian systems in old urban areas in the context of railway separation: a case study of Qingdao International Cruise Home Port start-up area and surrounding old urban areas

Lu Chunfang[1]　*Luan Yongpeng*[2]　*Dong Yunhao*[1]　*Peng Fangle*[1]
(1. Research Center for Underground Space, Tongji University, Shanghai 200092, China;
2. Qingdao City Civil Air Defense Construction Design Research Institute Corporation, Qingdao, Shandong 266061, China)

Abstract: Qingdao International Cruise Home Port start-up area is separated from surrounding old urban areas by a surface railway, making for the blocked surface pedestrian traffic. Thus, new requirements are put forward for the renewal of the external pedestrian system in old urban areas. In this paper, space syntax, a widely applied model in urban

planning and transportation network analysis, is used to evaluate pedestrian traffic system. Based on normalized angular integration (NAIN) and normalized angular choice (NACH), centralization and pass through capacity of pedestrian space can be obtained, which are used in the analysis of existing condition of railway separation between Cruise Home Port and old urban areas. Depending on the existing underground space, underground pedestrian streamlines and the present status of spatial difference computed by means of space syntax, planning schemes for external underground pedestrian traffic system are proposed. Finally, attraction potential of transit traffic and arrival-departure traffic are assessed with different schemes. This paper expects to be a reference by case analysis when it comes to the renewal and transformation of underground pedestrian passage of old urban areas which is confronted with railway separation. In addition, it is hoped that the application of space syntax in old urban areas in china provides lessons for the planning layout and evaluation of pedestrian system.
Keywords: underground pedestrian traffic; railway separation; old urban areas; space syntax

基于强度折减法的管廊双翼搭板控台背压实技术

黄俊文　廖　飞

（中建五局第三建设有限公司，湖南 长沙，410004）

摘　要： 有限空间台背回填压实度问题对市政道路工程的影响在短时间内很难被察觉，且极难通过有效的现场试验进行分析。强度折减法模拟回填料压实度变化的理念突破了传统的边坡稳定分析应用范畴，以回填料的强度、黏聚力为主控因素，建立了“管廊双翼搭板”地层-板相互作用力学模型，综合利用理论分析、数值模拟分析了“管廊双翼搭板”施工力学效应，揭示了板-土结构相互作用机理，提出了“管廊双翼搭板”设计计算方法。该方法避免了有限空间回填料压实度变化过程中路基与路面结构受力体系频繁转换问题，提高了结构质量与安全，降低了返工成本。

关键词： 有限空间；强度折减法；搭板；数值模拟

Compaction technology of console back of pipe gallery double wing approach slab based on strength reduction method

Huang Junwen　Liao Fei

(3rd Construction Co., Ltd. of China Construction 5th Engineering bureau, Changsha, Hunan 410004, China)

Abstract: The impact of the backfill compaction of the limited space on the municipal road engineering is difficult to detect in a short time, and it is extremely difficult to analyze through effective field tests. The concept of strength reduction method to simulate the change of backfill compaction degree breaks through the traditional application scope of slope stability analysis. With the strength and cohesive force of backfill as the main controlling factors, a"pipe corridor double-wing slab" stratum-slab interaction is established. The action mechanics model comprehensively utilizes theoretical analysis and numerical simulation to analyze the construction mechanics effects of the"pipe gallery double-wing slab", reveals the mechanism of the plate-soil structure interaction, and proposes the"pipe gallery double-wing slab" design calculation method. This method avoids the frequent conversion of the stress system of the roadbed and the pavement structure during the change of the compaction degree of the backfill in the limited space, improves the structural quality and safety, and reduces the rework cost.

Keywords: finite space; strength reduction method; approach slabt; numerical simulation

城市小客车专用地下道路经济型断面研究

郑智雄　杨子汉　张　立　李水生　何昌杰　罗杰峰
（中国建筑第五工程局有限公司，湖南 长沙，410007）

摘　要： 针对城市交通拥堵、土地资源紧张、城市财政能力有限、城市地下道路建设造价高的情况，提出了经济型的城市小客车专用地下道路的概念，并就不同驾驶模式下地下道路的行车道宽度、建筑限界高度、通行能力进行了研究分析。结果表明：1）城市小客车专用地下道路行车道宽度相对现行规范标准均可缩减 0.4～0.5m；2）仅通行小客车的情况下，建筑限界高度 2.5m 可以满足行车视距要求；3）智能驾驶模式下的理论通行能力是普通驾驶模式的 2 倍以上；4）当采取其他救援保障措施替代应急车道，同时缩减行车道宽度及建筑限界高度的情况下，城市地下道路断面面积可缩减 50％左右。研究成果为解决城市拥堵、减小城市基础设施建设财政压力提供了新思路。

关键词： 地下道路；车道宽度；建筑限界高度；智能驾驶；通行能力；经济断面

The research of economical cross-sections for the car dedicate urban underground road

Zheng Zhixiong　Yang Zihan　Zhang Li
Li Shuisheng　He Changjie　Luo Jiefeng
(China Construction Fifth Engineering Bureau Co., Ltd., Changsha, Hunan 410007, China)

Abstract: For urban traffic congestion, land resource constraints, the limited of urban fiscal and the high cost of urban underground road, the concept of economic car dedicate urban underground road have raised, meanwhile the carriageway width, building boundary height and traffic capacity of underground road under different driving mode was analyzed. The result shows that: 1) Compare with current criterion the carriageway width of urban underground road can decrease 0.4-0.5m; 2) In the case of cars only, the building boundary height of 2.5m can meet the driving sight distance requirements; 3) Theoretical capacity in intelligent driving mode is more than 2 times to the normal driving mode; 4) The cross-sectional area of urban underground roads can reduced about 50% when other rescue measures are taken to replace the emergency lane, simultaneously the width of the carriageway and the height of the building boundary are reduced. The research achievement are referenced a new way for solving urban congestion and reducing the financial pressure on urban infrastructure construction.

Keywords: underground road; carriageway width; construction clearance's height; intelligent driving; traffic capacity; economical cross-section

60m 级超大跨度暗挖地下洞室施工技术研究

王兴彬
（中铁隧道集团一处有限公司，重庆，401123）

摘　要：采用暗挖锚喷构筑法修建 60m 级超大跨度地下洞室的施工方法与施工工艺关乎洞室的施工安全、成洞结构安全和施工速度、工程造价。以跨度近 70m 跨度的某地下洞室工程为背景，通过工程类比初拟施工工法，采用 ANSYS 软件建立二维限元模型、FLAC3D 软件建立三维有限差分数值模型、3DEC 软件建立三维离散元数值模型进行施工过程模拟数值计算分析，同时分析了各工法的机械化适应性、施工效率和施工组织的难易程度。结果表明：三导洞洞口预留横向岩柱分部开挖、三导洞顺序分部开挖、双侧导洞预留横向岩柱间隔开挖三种方法都是安全的，具有可实施性；三个导洞洞口预留横向岩柱分部开挖法在施工组织、施工进度、大型机械化施工方面更有优势。工程实际采用三个导洞洞口预留横向岩柱分部开挖法，监测表明：洞库拱顶最大累计沉降值 22mm，净空收敛最大累计值 19mm，锚杆、锚索、喷混凝土内力均在允许范围内，保证了结构和施工安全；配置多臂凿岩台车、湿喷机械手等大型设备，实现了大型机械化施工，提高施工效率和降低劳动强度。

关键词：超大跨；暗挖地下洞室；三导洞预留岩柱开挖法；数值模拟；围岩控裂；施工监测

Research on construction technology of 60m class Super Large Span Underground Cavern

Wang Xingbin
（The First Construction Division Co.，Ltd. of China Railway Tunnel Group，Chongqing 401123，China）

Abstract：The construction method and technology of constructing 60m class super large span underground cavern with the method of underground excavation，anchor and shotcreting are related to the construction safety of cavern，the safety of tunnel structure，the construction speed and the project cost. Based on an underground cavern project with a span of nearly 70m，through engineering analogy with the preliminary construction method，the two-dimensional finite element model is established by ANSYS software，the three-dimensional finite difference numerical model is established by FLAC3D software，and the three-dimensional discrete element numerical model is established by 3DEC software to carry out the simulation numerical calculation and analysis of the construction process. At the same

time, the mechanization adaptability and the stability of each construction method are analyzed The results show that the three methods are safe and feasible, including the partial excavation of the reserved transverse rock column at the entrance of the three pilot tunnels, the sequential partial excavation of the three pilot tunnels, and the interval excavation of the reserved transverse rock column at the side pilot tunnels on both sides; The excavation method of reserving transverse rock column at three pilot tunnel openings has more advantages in construction organization, construction progress and large-scale mechanized construction. The monitoring results show that: the maximum cumulative settlement of vault is 22mm, the maximum cumulative value of headroom convergence is 19mm, the internal force of anchor bolt, anchor cable and shotcrete is within the allowable range, which ensures the safety of structure and construction; Large scale equipment such as multi arm rock drilling jumbo and wet spraying manipulator are equipped to realize large-scale mechanized construction, improve construction efficiency and reduce labor intensity.
Keywords: super large span; underground cavern excavation; three pilot tunnel reserved rock column excavation method; Numerical simulation; The surrounding rock controls the fracture; control of surrounding rock fracture

新加坡典型软土地层盾构穿河关键施工技术

肖　超[1,2]　喻畅英[1]　罗桂军[3]　刘　湛[2]
（中建五局土木工程有限公司，湖南 长沙 410004）

摘　要：新加坡地铁C715项目盾构区间下穿榜鹅河流，区间隧道底部距离河底仅5.61m，地层以粉质粒砂、黏土为主，盾构在该地层中施工易出现涌水涌砂、河堤沉降超限等问题，施工难度大，安全风险高。盾构机下穿河前，通过选择合理位置带压开仓检查刀具，穿越期间有效控制土仓压力、盾构掘进速度、渣土改良、出渣量，调整注浆压力、注浆量和盾尾油脂用量，同时进行同步注浆及沉降监测控制等措施，总结出一套符合新加坡业主要求，且在该地层土压平衡盾构下穿河流的关键施工技术，盾构安全、高效穿越榜鹅河，地表沉降及成型隧道质量均符合设计规范要求，可为类似工程施工提供参考。
关键词：土压平衡盾构；粉质粒砂；黏土；穿河

Key construction technology of shield driving through rivers in typical soft soil stratum of Singapore

Xiao Chao[1,2]　*Yu Changying*[1]　*Luo Guijun*[3]　*Liu Zhan*[2]
（CCFED Civil Engineering Co.，Ltd.，Changsha，Hunan 410004，China）

Abstract：The shield section of the Singapore MRT C715 project crosses the Punggol River. The bottom of the tunnel is only 5.61m from the bottom of the river. The stratum is dominated by silty sand with mixed clay. The construction of the shield in this stratum is prone to water gushing sand and river banks. Problems such as over-limit settlement make construction difficult and high safety risks. Before the shield machine crosses the river, check the tools by selecting a reasonable position and opening the warehouse under pressure. During the crossing, the pressure of the soil warehouse, the tunneling speed of the shield, the improvement of the slag, the amount of slag, and the adjustment of the grouting pressure, the amount of grouting and the shield are effectively controlled. The amount of tail grease, simultaneous grouting and settlement monitoring and control measures, summarized a set of key construction technologies that meet the requirements of the Singapore owner and that the earth pressure balance shield passes through the river in this formation. The shield passes through Punggol safely and efficiently. River, surface settlement and the quality of the formed tunnel meet the requirements of the design specification, which can provide a reference for the construction of similar projects.
Keywords：earth pressure balance shield；silt sand；clay；river crossing

城市高密度地区地下空间开发策略

乐迎春　金晓明
（广州建筑工程监理有限公司，广东 广州，510030）

摘　要：为解决城镇化发展过程中人口、城市、环境之间的矛盾，改善市民居住环境，需要加大对城市高密度地区地下空间的开发和利用，建立立体化城市空间发展模式，促进城市地上、地下之间的协调发展。城市高密度地区空间开发方式主要包括CBD地区空间一体化发展、新城中心地区地下空间一体化开发、基于旧城保护的地下空间开发和旧城高密度地区地下商业街开发等模式。高密度地区地下空间规划建设模式分为环形联结模式和脊轴带动模式。
关键词：城市；高密度地区；地下空间；开发

Development strategy of underground space in urban high density area

Yue Yingchun　Jin Xiaoming
(Guangzhou Construction Engineering Supervision Co.,
Ltd., Guangzhou, Guangdong 510030, China)

Abstract: in order to solve the contradiction among population, city and environment in the process of urbanization and improve the living environment of citizens, it is necessary to increase the development and utilization of underground space in urban high-density areas, establish a three-dimensional urban space development mode, and promote the coordinated development of urban above ground and underground. The spatial development modes of urban high-density areas mainly include the integrated development of CBD area, the integrated development of underground space in the central area of new city, the underground space development based on the protection of old city and the underground commercial street development in the high-density area of old city. The planning and construction mode of underground space in high-density area can be divided into ring connection mode and ridge axis driven mode.
Keywords: city; high density area; underground space; development

基于富水地层狭小空间的盾构双向始发技术研究

陈亚军　吕　涛　龙　彪　胡　敏
（中建五局土木工程有限公司，湖南 长沙，410004）

摘　要：依托长沙市万家丽路220kV电力隧道工程为工程背景，对富水地层狭小空间的盾构双向始发技术进行研究，详细介绍了一种先垂直分体后水平分体的双向始发技术。通过该技术的研究和应用，成功解决了地下水丰富和空间狭小的始发难题，提高了施工效率，为类似工程提供参考依据。
关键词：盾构法；双向始发；富水地层；狭小空间

Research on two-way launching technology of shield machine based on water-rich strata and small space

Chen Yajun　Lv Tao　Long Biao　Hu Min
（CCFEB Civil Engineering Co.，Ltd.，Changsha，Hunan 410004，China）

Abstract：Based on the engineering practice of Wanjiali road 220kV electric power tunnel engineering in Changsha，research on Two-way launching technology of shield machine in water-rich strata and small space，a two-way launching technology of vertical-separate first and horizontal-separate later is introduced in detail. Through research and application of this technology，to solve the shield machine launching problems of rich groundwater and small space. The construction efficiency is improved and it provides a reference for similar projects.
Keywords：shield method；two-way launching；water-rich strata；small space

专题八　城市防灾减灾

钢筋与 UHPC 粘结滑移性能及本构关系研究

李新星　周　泉　李水生
（中国建筑第五工程局有限公司，湖南 长沙，410000）

摘　要：基于钢筋开槽内贴应变片，通过 3 组锚固长度分别为 $3d$、$4d$ 和 $5d$ 的中心拉拔试件和 2 组保护层厚度为 25mm 和 45mm 的偏心拉拔试件，研究了不同锚固长度和保护层厚度的钢筋与 UHPC（超高性能混凝土）的粘结滑移性能。钢筋的锚固长度为 $3d$、保护层厚度为 25mm，试件破坏形式为钢筋拔出破坏；其余试件均发生钢筋拔断破坏。锚固长度越短，粘结应力越大，高应力区域较大，粘结滑移曲线更加饱满，靠近加载端位置的应力与加载端处相差较小；锚固段较长时，粘结应力基本不传递到自由端，粘结滑移曲线更加陡峭。保护层厚度越小的试件，粘结滑移曲线较保护层厚度较大的曲线更加饱满，高应力区域较大，粘结应力主要集中在加载端附近。通过位置函数 ψ（x）和平均粘结滑移关系 τ（s），建立了考虑粘结位置的钢筋与 UHPC 粘结滑移本构关系。
关键词：UHPC；钢筋；拉拔试验；粘结滑移；本构关系

Study on bond slip behavior and constitutive relationship between reinforcement and UHPC

Li Xinxing　Zhou Quan　Li Shuisheng
（China Construction Fifth Engineering Division Co.，Ltd.，
Changsha，Hunan 410000，China）

Abstract：The bond slip behavior of UHPC（ultra-high performance concrete）with different anchorage length and cover thickness was studied by using three groups of central pull-out specimens with anchorage length of $3d$，$4d$ *and* $5d$ and two groups of eccentric pull-out specimens with cover thickness of 25mm and 45mm，respectively. The anchorage length of reinforcement is $3d$，the thickness of protective layer is 25 mm，and the failure mode of specimen is pull-out failure; The rest of the specimens were broken. The shorter the anchorage length is，the larger the bond stress is，the larger the high stress area is，the fuller the bond slip curve is，and the difference between the stress near the loading end and that at the loading end is small; When the anchorage section is long，the bond stress is not transferred to the free end，and the bond slip curve is steeper. The smaller the thickness of the protective layer，the fuller the bond slip curve and the larger the high stress area. The bond stress is mainly concentrated near the loading end. By position function ψ（x）and average bond slip relationship τ（s）. The bond slip constitutive relationship between the reinforcement and UHPC considering the bond position is established.
Keywords：UHPC；reinforcement；pull-out test；bond slip；constitutive relation

高压细水雾灭火系统在地铁车站电气设备房间应用试验研究

杨　惠
（天津轨道交通集团有限公司，天津，300000）

摘　要： 地铁作为现代城市公共交通中的重要运输载体，其自动灭火系统多采用气体灭火系统，受其灭火机理限制，在有人电气房间尚未设置自动灭火系统。鉴于国内高压细水雾系统灭火实验对应用支撑不足，本文通过对高压细水雾灭火系统在地铁电气设备火灾扑灭能力、电气设备耐受时间和能力、电气设备恢复技术及恢复流程进行试验，验证高压细水雾自动灭火系统从灭火机理、灭火效能和安全性上可以替代常规的气体灭火系统。
关键词： 高压细水雾；地铁车站；电气设备房间

Experimental study on application of high-pressure water mist fire extinguishing system in electrical equipment room of subway station

Yang Hui
（Tianjin Rail Transit Co. , Ltd. , Tianjin 300000, China）

Abstract: As an important transport carrier in modern urban public transport, the automatic fire-extinguishing system of subway mostly adopts gas fire-extinguishing system. Due to the limitation of its fire-extinguishing mechanism, the automatic fire-extinguishing system has not been set up in the manned electrical room. In view of the lack of support for the application of domestic high-pressure water mist system fire-fighting experiments, this paper tests the fire-fighting ability of high-pressure water mist fire-extinguishing system in subway electrical equipment, electrical equipment endurance time and ability, electrical equipment recovery technology and recovery process, and verifies that the high-pressure water mist automatic fire-extinguishing system can replace the conventional fire-extinguishing mechanism, fire-fighting efficiency and safety gas fire extinguishing system.
Keywords: high-pressure water mist; subway station; electrical equipment room

南昌地铁盾构隧道整环结构三维数值分析

黄展军[1]　张琼方[2]

（1. 南昌轨道交通集团有限公司，江西 南昌，330000；

2. 中国电建集团华东勘测设计研究院有限公司，浙江 杭州，311122）

摘　要：结合南昌地铁盾构隧道工程实例，利用大型通用有限元软件 ABAQUS 建立地铁盾构隧道整环衬砌结构的三维有限元模型。衬砌及螺栓等采用三维实体单元，混凝土采用弹性损伤模型，螺栓和钢筋采用双折线模型，分析了计算分析衬砌环的荷载—变形关系曲线，管片发生收敛位移时螺栓、钢筋和混凝土的应力应变变化规律。

关键词：地铁盾构隧道；有限元分析；应力应变

Abstract: Based on Nanchang Metro shield tunnel, the three-dimensional finite element model of the whole ring lining structure of the metro shield tunnel is established by using the large-scale general finite element software ABAQUS. The three-dimensional solid element is used for the lining and bolts, the elastic damage model is used for the concrete, and the double line model is used for the bolts and steel. The stress and strain of bolt, steel bar and concrete is analyzed when the segment converges.

Keywords: shield tunnel; finite element analysis; stress and strain

关于气象观测的降雨特征与城市排水体制的关联研究

曾庆红
（江西省建筑设计研究总院集团有限公司，江西 南昌，330046）

摘　要： 本文从建立排水分流制城市的排水系统和水环境污染模型入手，论述了排水分流制中的雨水排水管渠存在日常污废水排出的客观事实。通过城市任意年的日降雨量、降雨历时的气象观测记录与城市不同的排水体制的污水溢流百分率的关联计算，比较计算结果，找出不同排水体制的优劣性。
关键词： 城市水环境；降雨量；降雨历时；污水溢流量计算；污水溢流百分率计算；排水体制

Study on the relationship between rainfall characteristics of meteorological observation and urban drainage system

Zeng Qinghong
(Jiangxi Architectural Design Research Institute Group Co.，
Ltd.，Nanchang，Jiangxi 330046，China)

Abstract：Starting with the establishment of drainage system and water environment pollution model，this paper discusses the objective fact that there is daily sewage discharge in the rain water drainage pipe and canal in the drainage diversion system. Through the correlation calculation between the daily rainfall and rainfall duration of any year in the city and the percentage of sewage overflow of different drainage systems in the city，the calculation results are compared to find out the advantages and disadvantages of different drainage systems.
Keywords：urban water environment；rainfall；rainfall duration；sewage overflow calculation；sewage overflow percentage calculation；drainage system

黄土区域地铁降水引发高层建筑过度沉降的处理技术

安刘生
（中国建筑第五工程局有限公司，湖南 长沙，410004）

摘　要：施工降水对周边建构筑物的影响一直是工程界学者们探讨的重要课题，随着地铁工程的迅速开展，施工降水引发的高层建筑沉降事故层出不穷。我国北方大部分区域位于黄土区，黄土特有的高湿陷、高压缩特性，使施工降水对周边建筑物的影响分析及过度沉降后的处理技术更趋复杂。本文通过对西安地铁八号线施工降水引发的高层住宅过度沉降处理技术的总结，为类似黄土区域的沉降处理技术提供借鉴。

关键词：地铁；黄土；降水；高层建筑；沉降

Treatment technology of high-rise building excessive subsidence caused by metro rainfall in loess area

An liusheng
（China Construction Fifth Engineering Bureau Co.，Ltd.，
Changsha，Hunan 410004，China）

Abstract：The impact of construction precipitation on surrounding structures has always been an important topic discussed by scholars in the engineering field. With the rapid development of subway projects，high-rise building settlement accidents caused by construction precipitation have emerged endlessly. Most areas in northern my country are located in the loess area. The unique characteristics of high collapsibility and high compression of loess make the analysis of the impact of construction precipitation on surrounding buildings and the treatment technology after excessive settlement more complicated. This article summarizes the treatment technology for the excessive settlement of high-rise residential buildings caused by the precipitation of the construction of Xi′an Metro Line 8 to provide a reference for the settlement treatment technology of similar loess areas.

Keywords：subway；loess；precipitation；high-rise building；settlement

应对气象灾害的寒地乡镇医疗机构空间保障设计研究

王　田　张姗姗　余振凡
（哈尔滨工业大学，黑龙江 哈尔滨，150001）

摘　要：我国高纬度寒冷地区面积广阔，气候条件恶劣，以低温冷冻害、大风、冰雹、洪涝寒潮和雪灾为代表的气象灾害频发，成为严重制约我国寒地乡镇卫生院的发展、威胁乡镇居民的公共健康安全的重要因素。此外，我国乡镇经济文化水平与基础设施较差，导致寒地乡镇医疗发展滞后，且现有的建筑应灾空间设计研究缺乏针对寒地乡镇卫生院这一特定对象的针对性研究成果。本研究采用实地调研、文献调研与类比调研相结合的调研方法提出寒地乡镇医疗机构的应灾设计对策，医疗机构的空间保障设计从规划布局设计、内部空间设计等方面的提升，满足了气象灾害下寒地乡镇卫生院与乡镇居民的特殊需求，对寒地乡镇卫生院的建设及发展具有理论意义和实践意义。
关键词：寒地；乡镇医疗；气象灾害；应灾设计

Study on the spatial security system of township medical institutions in cold regions responding to meteorological disaster

WangTian　Zhang Shanshan　Yu Zhenfan
(Harbin Institute of Technology, Harbin, Heilongjiang 150001, China)

Abstract: China's high latitude cold region has a vast territory and bad climate conditions. The frequent meteorological disasters, such as low temperature, cold damage, gale, hail, flood, cold wave and snow disaster, have seriously restricted the development of township health centers in cold regions and threatened the public health and safety of township residents. In addition, the poor economic and cultural level and infrastructure of villages and towns in China lead to the lag of medical development in cold villages and towns, and the existing research on building disaster response space design lacks targeted research results for the specific object of cold villages and towns health centers. In this study, field research, literature research and analogy research methods were used to put forward the disaster response design countermeasures of township medical institutions in cold regions, and the spatial security system of medical institutions was established. The planning layout design and disaster response system design were improved to meet the special needs of township health centers and residents in cold regions under meteorological disasters, It has theoretical and practical significance for the construction and development of township health centers in cold regions.
Keywords: cold area; township medical service; meteorological disasters; disaster response design

基于“海绵城市”理念的大型城市林带雨水综合处理系统研究

权利军　黄　蜀　刘　科
（中建五局第三建有限公司西北分公司，陕西 西安，710000）

摘　要：海绵城市是一种与可持续发展战略高度契合的理念，是一种以雨水综合管控为出发点的城市建设新模式，是统筹解决水资源、水环境、水安全等水系统问题的重要措施和手段。本文依托西安幸福林带建设工程，对“海绵城市”理念下的城市大型林带雨水综合处理系统展开研究，浅析相关设计策略、设计原则及设计方案，以期为后续工程提供借鉴。
关键词：海绵城市；幸福林带；水资源

Research on the comprehensive rainwater treatment system of large urban forest belt based on the concept of "sponge city"

Quan Lijun　Huang Shu　Liu Ke
(China Construction Fifth Bureau Third Construction Co., Ltd.
Northwest Branch, Xi'an, Shanxi 710000, China)

Abstract: Sponge city is a concept of high perseverance with sustainable development strategy, a new model of urban construction with comprehensive rainwater control as the starting point, and an important measure and means to comprehensively solve water resources, water environment and water safety. Relying on the construction project of Xi'an happiness forest belt, this paper studies the comprehensive rainwater treatment system of large urban forest belt under the concept of "sponge city", and analyzes the relevant design strategies, design principles and design scheme, in order to provide reference for subsequent projects.
Keywords: sponge city; happy forest belt; water resources

协同应变 有机共生
——城市空间防灾能力提升策略

王　田　张姗姗
（哈尔滨工业大学，黑龙江 哈尔滨，150001）

摘　要：针对公共卫生事件爆发的偶然性与必然性，把公共危机的普通性与特殊性有机结合，从公共安全的角度入手解析城市空间秩序在突发公共卫生事件应对流程中的建构，以健康导向空间规划为原则，提出提升城市空间防灾能力的规划策略，对协同应对突发公共卫生事件具有理论意义和现实应用的紧迫感。
关键词：公共卫生事件；城市空间；防灾能力；防灾空间

Synergistic strain, Organic symbiosis: Strategies for improving urban space disaster prevention capacity

Wang Tian　Zhang Shanshan
(Harbin Institute of Technology, Harbin, Heilongjiang 150001, China)

Abstract: Regarding the contingency and inevitability of the outbreak of public health events, this article combines the generality and particularity of public crisis, and starts from the perspective of public safety to analyze the construction of urban spatial order in the response process of public health emergencies. Based on the principle of health-oriented spatial planning, it puts forward the strategy to improve the capacity of urban space disaster prevention . The results of this article have theoretical significance for a collaborative response to public health emergencies and could be put into the application as soon as possible due to the urgency of the actual application.
Keywords: public health events; urban space; disaster prevention capacity; disaster prevention space

沙河涌流域南方医院段新建雨水渠箱排涝施工及其必要性

傅海森

（广州建筑工程监理有限公司，广东 广州，510030）

摘　要：通过工程介绍沙河涌流域南方医院段现状渠箱过水能力不足，为确定沙河涌流域新建雨水渠箱排涝施工方式的合理性，采用地下连续墙导墙施工、钢筋笼制作及吊装施工、地下连续墙混凝土浇筑施工、冠梁及支撑梁施工、钢制渠箱深基坑开挖、雨水渠箱底板及顶板施工、钢制渠箱主体结构及污水检查井整体混凝土浇筑施工以及钢制渠箱回填及路面修复等工艺，所需设备有长短臂挖掘机、抓槽机、吊机等。新建的雨水渠箱具有分流功能、收水能力良好，性能稳定，基本解决现状旧雨水渠箱过水能力不足、南方医科大学南方医院及广州大道北周围水浸点缺少雨水收水措施或收水口堵塞严重问题。结合本工程探讨了广州市黑臭河涌排水体制的选择及设计存在的问题，并对广州市黑臭河涌排水系统建设提出了一些建议。

关键词：地下连续墙；雨水渠箱；排水体制

Construction and necessity of new stormwater drainage box drainage in south hospital section of shaheyong drainage basin

Fu Haisen

(Guangzhou Construction Engineering Supervision Co., Ltd., Guangzhou, Guangdong 510030, China)

Abstract: In order to determine the rationality of the construction method of new stormwater drainage box drainage in Shaheyong drainage basin, this paper introduces the current situation of the southern hospital section of Shaheyong drainage basin with insufficient water passing capacity. Guide wall in underground continuous wall construction, reinforcing cage and lifting construction, concrete pouring construction of underground continuous wall, crown, steel channel box beam and support beam construction of deep foundation pit excavation, storm drains, floor and roof construction, steel channel box main body structure and sewage well overall construction and concrete casting steel ditch backfilling and road repair process, The required equipment includes long and short arm excavator, grooving machine, crane and so on. The newly built stormwater drainage box has the function of diversion, good water collecting capacity and stable performance, which basically solves

the problems of insufficient water passing capacity of the old stormwater drainage box, the lack of rainwater collecting measures or the blockage of water collecting ports in the flooding points around Nanfang Hospital of Southern Medical University and the north of Guangzhou Avenue. Based on the project, the problems in the selection and design of drainage system of black and odorous river in Guangzhou are discussed, and some suggestions for the construction of black and odorous river drainage system in Guangzhou are put forward.
Keywords: diaphragm wall; the rain water tank; the drainage system

专题九　土木工程高质量发展

基于韧性城市理念的城市轨道交通设计探讨

凌景文
（中铁第四勘察设计院集团有限公司，湖北 武汉，430063）

摘　要：根据韧性城市的理念框架，总结梳理了韧性城市的概念和特性。结合设计任务和内容，分析了轨道交通设计韧性的框架体系，从设计实践出发探讨了城市轨道交通在交通衔接、灾害防范、应急处理、系统选型等方面的韧性理念应用。
关键词：韧性城市；交通衔接；轨道交通；框架体系

Discussion of rail transit design based on the concept of resilient city

Ling Jingwen
（China Railway Siyuan Survey and Design Group Co.，Ltd.，Wuhan，Hubei 430063，China）

Abstract：According to the concept framework of resilient city，this paper summarizes the concept and characteristics of resilient city. Combined with the Rail Transit design tasks and contents，this paper analyzes the framework system of rail transit design resilience，and from the design practice，discusses the application of urban rail transit resilience concept in traffic connection，disaster prevention，emergency treatment，system selection，etc.
Keywords：resilient city；traffic connection；rail transit framework system

国家会展中心（天津）二期工程交通连廊内多类桁架施工受力分析与研究

王俊伟　李　晨　周　磊　贾聪亮　王　刚
（中国建筑第八工程局有限公司，上海，200135）

摘　要：以国家会展中心（天津）二期工程大跨度交通连廊为背景，针对结构内置不同位置组合桁架体系进行施工过程受力特性研究。本文介绍了钢桁架结构安装过程中需要考虑的主要因素，对桁架安装不同阶段进行有限元受力分析，总结出桁架结构在不同安装不同种的静力特性和内力分布状态等特点。通过对比分析得出不同施工阶段分析和常规简化分析结果的差异及结构的静力特性，能为类似工程钢桁架结构安装提供借鉴和参考。
关键词：钢结构桁架；施工阶段分析；受力状态

Construction stress analysis and research of multiple truss in traffic corridor of National Convention and Exhibition Center（Tianjin）phaseII project

Wang Junwei　Li Chen　Zhou Lei　Jia Congliang　Wang Gang
（China Construction Eighth Engineering Division Co.，Ltd.，Shanghai 200135，China）

Abstract：Based on the large-span traffic corridor of National Convention and Exhibition Center（Tianjin）phase II project，the mechanical characteristics of the built-in composite truss system at different positions during construction are studied. This paper introduces the main factors that need to be considered in the installation process of steel truss structure，analyzes the stress of truss in different stages of installation by finite element method，and summarizes the static characteristics and internal force distribution of truss structure in different installation stages. Through the comparative analysis，the differences between the analysis results of different construction stages and the conventional simplified analysis results and the static characteristics of the structure are obtained，which can provide reference for the installation of similar steel truss structures.
Keywords：steel truss；analysis of construction stage；stress state

可调节操作平台在钢结构工程中的应用

韩宗辉　张俊宏　刘成成
（中国建筑第八工程局有限公司钢结构工程公司，上海，200000）

摘　要： 在超高层高结构安装过程中，目前大部分的操作平台结构简单，安全性能差、周转效率差、重复利用率低，给钢结构施工安全和效率造成严重影响。可调节操作平台从根本上解决了以上问题，推进施工现场防护设施向标准化的方向发展。
关键词： 超高层钢结构；可调节操作平台；安全防护

Application of adjustable operation platform in steel structure engineering

Han Zonghui　Zhang Junhong　Liu Chengcheng
(China Construction Eighth Engineering Division Corp.,
Ltd Steel Structure Engineering Company, Shanghai 200000, China)

Abstract: In the process of super high-rise steel structure installation, most of the operation platforms have simple structure, poor safety performance, poor turnover efficiency and low reuse rate, which seriously affect the safety and efficiency of steel structure construction. The adjustable operation platform fundamentally solves the above problems and promotes the development of construction site protection facilities to the direction of standardization.
Keywords: super high rise steel structure; adjustable operating platform; safety protection

真空预压变形理论分析及应用

胡利文　董志良　王　婧　梁小丛
（中交四航工程研究院有限公司，广东 广州，510230）

摘　要：从调研真空预压变形分析存在的问题出发，基于应力应变关系研究了真空预压加固机理，采用有限元法分析了真空预压不同位置土单元的应力路径，进而按一维固结和各向等压固结应力路径依据修正剑桥模型得出不同类型真空预压变形计算方法和弹塑性状态下软土的理论静止土压力系数 K_{0MC}。基于理论分析，对大面积真空预压软基沉降进行验证，通过案例分析了无砂垫层增压式真空预压及水下真空预压等的创新应用。研究表明，现场真空预压应力路径随土单元位置而变，不能保证各向等压固结。真空预压土体沉降除与土特性有关外，还与加固区尺度有关，对不同面积加固区提供了计算建议方法，并且当加固区宽度是加固深度的4～5倍时，可采用一维计算公式计算沉降；根据理论创新应用，加强工艺改进，可有效发展和利用综合真空预压技术加固软土地基。
关键词：真空预压；加固机理；应力路径；变形；应用

Deformation mechanism of soft soil subjected to vacuum preloading and its application

Hu Liwen　Dong Zhiliang　Wang Jin　Liang Xiaocong
(CCCC Fourth Harbor Engineering Institute Co. , Ltd. ,
Guangzhou, Guangdong 510230, China)

Abstract: To understand the deformation mechanism of soft soil subjected to vacuum preloading, the disagreement on settlement yielded by vacuum preloading is investigated and verified by the different stress paths of soil elements at different locations by employing finite element method. Based on modified Cam-clay model, predictions of settlement of soil under various types of vacuum preloading are deduced according to one dimensional consolidation and isotropic loading consolidation, as well as the theoretical ratio of earth pressure at rest K_{0MC} for the soil in elastic-plastic state. The settlement of soil in large-area of treated zone is calculated using the deduced prediction equations and compared with the measured. Finally, either enhanced vacuum preloading assisted by air pressure without sand blanket or underwater vacuum preloading are presented to rethink the deformation mechanism of soil subjected to the vacuum type preloading. It is founded that stress path of soil element varies with locations for vacuum preloading in site, and the isotropic loading condi-

tion is not guaranteed。The magnitude of settlement in the center of treated zone is related to the area scale of treated zone including soil properties, and a discount factor is applied to predict settlement for narrow breadth of treated zone, while one dimensional loading condition may be suggested for treated zone with breadth larger than 4-5 times treated depth. Based on the deformation expression, composite vacuum type ground improvement can be innovated for a more effective way.
Keywords: vacuum preloading; mechanism of ground improvement; stress path; deformation; application

某超高层建筑异形钢柱加工工艺浅析

杨　彬　董　涛　艾　伟　徐　铭　马晓伟　董龙超
（中国建筑第八工程局钢结构工程公司，上海，200120）

摘　要：超高层钢结构具有自重轻、强度高、施工快捷可实现结构稳定、造型复杂等特点。近几十年来在中国地区得到了广泛的应用，是一项成熟的建筑结构体系，非常适合我国未来的发展方向。某超高层建筑为双曲面钢结构，其类型主要有异形圆管柱、异形桁架等。在某超高层钢结构建筑中，钢柱内环板数量多、材料利用率低，环形板件坡口质量差、焊缝合格率低，圆管柱外伸节点多、制作难度大。在制作过程中通过钢柱内环板拼接、变径圆管预制、圆管椭圆度校正、虚拟预拼装等新工艺指导下进行施工，利用 BIM 模型检验拼装尺寸的检测方法提升空间点的尺寸精度，在工程的整体质量提升情况下，实现“鲁班奖”工程的目标。在工程的施工过程中通过新技术总结出加工难点并制定有效解决措施、提升制作效率、保障加工精度经验，可在曲面异形钢结构件的制作及检验方面推广和使用。

关键词：钢结构；环板拼接；柱梁节点整体制作；圆管椭圆度；异性钢牛腿组装

Analysis on the processing technology of special-shaped steel column in a super high-rise building

Yang Bin　Dong Tao　Ai Wei　Xu Ming　Ma Xiaowei　Dong Longchao
(China Construction Eighth Engineering Bureau Steel Structure Engineering Company , Shanghai 200120, China)

Abstract: The super high-rise steel structure has the characteristics of light weight, high strength, fast construction, stable structure and complex shape. In recent decades, it has been widely used in China. It is a mature building structure system, which is very suitable for the future development direction of our country. A super high-rise building is a hyperboloid steel structure, its main types are special-shaped circular pipe column, special-shaped truss and so on. In a super high-rise steel structure building, it is found that there are many annular plates in the column, the utilization rate of materials is low, the groove quality of annular plates is poor, the qualification rate of welds is low, and there are many protruding joints of the circular pipe column, making it difficult. In the manufacturing process, the construction is conducted under the guidance of new technologies such as steel column inner ring plate splicing, variable diameter circular pipe prefabrication, circular pipe ovality correction, virtual pre assembly, etc. , and the measurement method of BIM

model inspection assembly size is used to improve the dimensional accuracy of space points, so as to achieve the goal of " Luban Prize" project under the overall quality improvement of the project. In the construction process of the project, the processing difficulties are summed up through new technology, and effective measures are formulated to improve the production efficiency and guarantee the processing accuracy experience, which can be popularized and used in the production and inspection of curved shaped steel structures.
Keywords: steel structure; ring plate splicing; overall fabrication of column beam joints; ovality of circular pipe; assembly of opposite steel corbel

浅析降低大跨度悬挑屋面超厚板钢结构变形率

董　涛　艾　伟　马晓伟　杨　彬　高月健　徐　铭
（中国建筑第八工程局钢结构工程公司，上海，200120）

摘　要： 在钢结构中，如何控制悬挑钢结构安装变形是一大难点，本文从制作、运输、吊装、测量、焊接等每道工序，对大跨度悬挑屋面超厚板钢结构变形问题进行了的深入研究，从中找出了关键因素并加以控制，保证了钢屋面的施工质量。本文对所取得的成果及所采用的对策、措施形成技术总结，有利于同行业类似工程交流和借鉴，也为今后类似工程的施工积累了施工管理和技术经验。
关键词： 钢结构；控制；悬挑；大跨度；超厚板；变形

Analysis on bending assembly of arc steel beam without diaphragm

Dong Tao　Ai Wei　Ma Xiaowei　Yang Bin　Gao Yuejian　Xu Ming
(China Construction Eighth Engineering Bureau Steel Structure
Engineering Company, Shanghai 200120, China)

Abstract: In steel structure, how to control the installation deformation of overhanging steel structure is a big difficulty. In this paper, from the production, transportation, hoisting, measurement, welding and other processes, the deformation of the steel structure of large-span cantilevered roof is deeply studied, and the key factors are found out and controlled to ensure the construction quality of the steel roof The technical summary of countermeasures and measures is conducive to the exchange and reference of similar projects in the same industry, and also accumulates construction management and technical experience for the construction of similar projects in the future.
Keywords: steel structure; control; overhanging; large span; ultra thick plate; deformation

浅析无隔板式弧形钢梁的弯曲组立

董　涛　杨　彬　艾　伟　马晓伟　徐　铭　高月健
（中国建筑第八工程局钢结构工程公司，上海，200120）

摘　要：在钢结构中，无隔板式弧形钢梁的制作是一个难点，其中圆弧弧度放样组对过程最为重要，例如圆弧钢梁截面地样无法对钢梁组立、外委费用较高等问题。以往加工厂采用手工组立，费时、费力，精度难以掌控。本文通过对原有工程的总结，采用模具组对方式组立，运用此方式进行加工制作，可以有效减少工程加工成本，缩短制作工期，提高制作效率、制作精度及经济效益，适用于大批量生产，具有较好的市场应用前景。
关键词：钢结构；圆弧；无隔板；钢梁；弯曲；组立

Analysis on bending assembly of arc steel beam without diaphragm

Dong Tao　Yang Bin　Ai Wei　Ma Xiaowei　Xu Ming　Gao Yuejian
(China Construction Eighth Engineering Bureau Steel Structure Engineering Company, Shanghai 200120, China)

Abstract: In the steel structure, it is a difficult point to make circular arc section steel without partition board. Among them, the setting out group of arc radian is the most important to the process. For example, the sample of arc section steel can not assemble the section steel, and the outsourcing cost is high. In the past, manual assembly was used in processing plants, which was time-consuming and laborious, and the accuracy was difficult to control. In this paper, through the summary of the original project, using the mold assembly method assembly, using this way for processing, can effectively reduce the engineering processing cost, shorten the production period, improve the production efficiency and production accuracy, improve the economic benefits of both aspects, suitable for mass production, has a good market application prospects.
Keywords: steel structure; circular arc; no partition; steel beam; bend; assemblage

铁路客站钢屋盖结构抗连续倒塌分析

姜 锐
(国铁集团 工程管理中心，北京，100844)

摘 要：大型铁路客站在遭受偶然事件引发局部破坏从而导致连续倒塌时，会造成大量的人员伤亡和经济损失，应对其抗连续倒塌能力进行分析研究。以厦门北站钢屋盖结构体系为研究对象，先用敏感性分析确定结构的关键构件，进而利用拆除构件法，对关键构件拆除后的结构进行线性静力分析和非线性动力分析。本文研究结果表明：柱敏感性系数与柱的受荷面积直接相关；拆除关键构件 XZ1 后，无论线性静力分析还是非线性动力分析，剩余结构的最不利杆件是一致的，均为与其相连的圆管斜拉杆。所不同的是，线性静力分析偏保守，在应力超过抗拉强度最小值后，判断杆件失效；而非线性分析显示，虽然该杆件出现塑性，但仍可继续使用；两种分析结果均表明，即使 XZ1 失效，破坏也仅限于柱节点支撑杆件，并没有向屋盖延伸，说明该体系具有比较好的抗连续倒塌能力。

关键词：铁路站房；连续倒塌；拆除构件法；关键构件；敏感性分析

Anti-progressive collapse analysis of steel structure in railway station

Jiang Rui
(Engineering Management Center of China Railway, Beijing 100844, China)

Abstract: The progressive collapse of large railway station may be initiated by the local failures in case of unexpected loads, which will cause severe casualties and economic losses. It is necessary to carry out research work of the progressive collapse of large railway station. The linear static analysis and non-linear dynamic analysis were conducted for steel structure of Xiamen North railway station employing alternative path method. The research result shows: the sensitivity coefficient of column is related to the loading-carrying area; the most unfavorable component is the diagonal member connected to the invalid column XZ1; the linear static analysis is more conservative than non-linear dynamic analysis; the damage is confined in a limited area and the roof structure has ability to resist progressive collapse.

Keywords: railway station; progressive collapse; alternate path method; key component; sensitivity analysis

商业大跨度采光顶施工平台技术应用

伍　鸽　黄　俊　欧阳喆　何　斌　曹鸿皓
（中国建筑第五工程局有限公司，湖南　长沙，410007）

摘　要：株洲万达广场项目采光顶数量多、面积大、造型各异、施工难度大，本文采用在采光顶中搭设钢丝绳软平台和移动平台的方法，分别在圆形采光顶中搭设钢丝绳软平台和在方形采光顶中搭设移动操作平台，相比于采用传统的方法从底部搭设满堂架，此方案不仅可以节约大量材料、人工成本，还为地面工程、装饰工程提供了提前进场施工的条件，对工期与成本节约都有着重要的作用。
关键词：采光顶；大跨度；操作平台；施工技术

Application of commercial large span daylighting roof construction platform technology

Wu Ge　Huang Jun　Ouyang Zhe　He Bin　Cao Honghao
(China Construction Fifth Engineering Bureau Co. , Ltd. ,
Changsha, Hunan 410007, China)

Abstract: Zhuzhou Wanda Plaza project has a large number of lighting roofs, large areas and different shapes, making construction difficult. This paper adopts the method of installing a soft platform and mobile platform for wire rope in the lighting roof, and a mobile operation platform in the round lighting roof, respectively. Compared with the traditional method of building a full frame from the bottom, this scheme can not only save a lot of materials and labor costs, but also provides conditions for ground engineering and decoration projects to carry forward field construction, which has a positive impact on the construction period and cost saving.
Keywords: daylighting roof; large span; construction platform; engineering technology

准分布式光纤光栅传感器预应力监测性能研究

沈全喜[1,2] 张贺丽[3] 刘丰荣[3] 覃荷瑛[2,4] 朱万旭[2,4]
（1. 桂林理工大学 地球科学学院，广西 桂林，541006；
2. 广西智慧结构材料工程研究中心，广西 桂林，541004；
3. 广西汉西鸣科技有限责任公司，广西 桂林，541004；
4. 桂林理工大学 土木与建筑工程学院，广西 桂林，541004）

摘　要：钢绞线的应力监测是预应力结构安全性评估的重要方法，文章介绍了一种用于监测七丝钢绞线应力的凹槽预压封装准分布式光纤光栅传感器监测法，以解决光纤光栅传感器易破断和监测量程不够的技术难题。分析了光纤光栅传感器波长与应变的理论关系，并对三根自感知钢绞线进行重复张拉试验。试验结果表明，三根自感知钢绞线监测数据均有良好的线性度，应变灵敏度约为 0.00119nm/$\mu\varepsilon$，应变监测量程达到 7619$\mu\varepsilon$。利用上述试件制作三根预应力分别为 0.45f_{ptk}、0.65f_{ptk}、0.75f_{ptk}的无粘结预应力活性粉末混凝土梁，通过加载试验，研究发现，准分布式光纤光栅传感器能够对梁的开裂和受拉钢筋屈服进行准确识别，随着预应力增大，梁的开裂荷载与钢筋屈服时的荷载也随之增大。

关键词：准分布式光纤光栅传感器；自感知钢绞线；预应力监测；RPC 梁

Research on prestress monitoring performance of quasi-distributed fiber bragg grating sensor

Shen Quanxi [1,2] *Zhang Heli*[3] *Liu Fengrong*[3] *Qin Heying*[2,4] *Zhu Wanxu* [2,4]
(1. College of Earth Sciences, Guilin University of Technology, Guilin, Guangxi 541006, China; 2. Guangxi Intelligent Structure Materials Engineering Research Center, Guilin, 541004, China; 3. Guangxi hanximing Technology Co., Ltd., Guilin, Guangxi 541004, China; 4. College of Civil Engineering and Architecture, Guilin University of Technology, Guilin, Guangxi 541004, China)

Abstract: Stress monitoring of steel strand is an important method for the safety evaluation of prestressed structure. This paper introduced a kind of groove preloading package quasi-distributed fiber Bragg grating sensor monitoring method for the stress monitoring of seven-wire steel strand, so as to solve the technical problems of fiber bragg grating sensor easy to break and insufficient monitoring range. The theoretical relationship between wavelength and strain of the fiber grating sensor was analyzed, and the repeated tensile test was carried out on three self-sensing steel strands. The test results show that the monitoring data of three self-sensing steel strands have good linearity. Moreover, the strain sensitivity

is about 0. 00119nm/$\mu\varepsilon$, and the strain monitoring range reaches 7619$\mu\varepsilon$. Three unbonded prestressed reactive powder concrete beams with prestress of 0. 45 f_{ptk}, 0. 65 f_{ptk} and 0. 75 f_{ptk} were fabricated using the above specimens. Through the loading test, the study found that the quasi-distributed fiber Bragg grating sensor can accurately identify the cracking of the beam and the yield of the tensile reinforcement. With the increase of prestress, the cracking load of the beam and the yield load of the reinforcement also increase.
Keywords: quasi-distributed fiber grating sensor; self-sensing steel strand; prestress monitoring; RPC beam

考虑多因素的城市轨道交通短时客流深度学习预测

王子甲[1] 陈志翔[1] 陈雍君[2] 申晓鹏[3]
（1. 北京交通大学 土木建筑工程学院，北京，100044；
2. 北京建筑大学，北京，100044；
3. 中国国际工程咨询有限公司，北京，100044）

摘 要：小时粒度的短时客流预测，对城市轨道交通精准投放运力、预判早晚高峰大客流、采取限流等组织措施有重要意义。然而客流生成的机理复杂，影响因素众多，周期性、随机性和波动性兼有。既有研究在考虑因素、时间跨度、模型精度等方面尚存在问题。为充分挖掘城市轨道交通短时客流波动的内在机理，提高预测精度。本文基于深度学习方法，考虑长期、细粒度的时空天气数据，日期等因素，提出了一种考虑多因素的长短期记忆神经网络模型（MUL-LSTM）。利用 2019 年北京地铁 1 号线八宝山站的全年小时粒度进站客流及全年的小时粒度气象台站监测数据并结合日期属性数据，应用所建立的模型并标定了参数。结果表明，该模型最终的预测平均误差为 6.99%，具有较好的预测精度。

关键词：城市轨道交通；短时客流预测；多因素；深度学习

Short-term passenger flow forecasting in urban rail transit considering multiple factors based on deep learning method

Wang Zijia[1] *Chen Zhixiang*[1] *Chen Yongjun*[2] *Shen Xiaopeng*[3]
（1. Beijing Jiaotong university，School of civil engineering，Beijing 100044，China；
2. Beijing Jianzhu university，Beijing 100044，China；
3. China international engineering consulting corporation，Beijing 100044，China）

Abstract：The short-term passenger flow forecasting of hourly particle size is of great significance for the precise delivery of urban rail transit capacity，the prediction of large passenger flow，the adoption of current restriction and other organizational measures. However，the mechanism of passenger flow generation is complex，and there are many influencing factors，including periodicity，randomness and fluctuation. There are still some problems in previous studies，such as considerations，time span，model precision and so on. In order to fully explore the internal mechanism of the short-term passenger flow fluctuation of urban rail transit and improve the accuracy of prediction model. In this paper，a multiply-factors neural network model for long and short term memory（MU-LSTM）is proposed based

on the deep learning method, taking into account the long-term, fine-grained spatial and temporal weather data, date and other factors. Based on the annual hourly granularity of passenger flow at Ba bao shan Station of Beijing Metro Line 1 in 2019 and the annual hourly granularity of meteorological observatory and station monitoring data combined with date attribute data, the model and parameters are calibrated. The results show that the final prediction average error of the model is 6.99% and has a good prediction accuracy.
Keywords: urban rail transit; short-term passenger flow forecast; multiple factors; deep learning

预应力碳纤维板张弦梁组合加固构件的受力性能研究

黄伟哲　李悦峰　韩洪鹏　王保栋
（中建八局发展建设分公司，山东 青岛，266100）

摘　要：将传统的体外张弦梁组合结构的体外拉索用碳纤维板进行替换，得到体外预应力碳纤维板张弦梁组合结构，借用这种张弦梁结构的受力原理，本文在此基础上考虑将该技术应用于对桥梁结构进行加固，因此提出了预应力碳纤维板张弦梁组合体系。相比于传统的体表有粘结预应力碳纤维板加固，本技术的碳纤维板离被加固结构的形心更远，相同的预应力产生的偏心弯矩更大，预期加固效果会更显著。当然，该技术需要的空间更大，在外观上会略逊于常规的体表有粘结预应力碳板加固，但是该技术对于特殊条件下的桥梁加固，比如某些山区桥梁（桥下空间较大，且通航通车净空足够的前提下）有很好的工程应用价值，因此有必要开展预应力碳纤维板张弦梁组合加固技术的探索研究。

关键词：预应力碳纤维板；张弦梁组合结构；桥梁加固；理论分析

Study on the mechanical performance of the prestressed carbon fiber board string beam composite strengthening member

Huang Weizhe　Li Yuefeng　Han Hongpeng　Wang Baodong
(China Construction Eighth Engineering Bureau Co., Ltd.,
Qingdao, Shandong 266100, China)

Abstract: The external cable of the traditional external stringed beam composite structure is replaced with carbon fiber board to obtain the externally prestressed carbon fiber board stringed beam composite structure. Borrowing the force principle of this stringed beam structure, this paper considers applying the technology to bridges on this basis. The structure is reinforced, so a combination system of prestressed carbon fiber board string beams is proposed. Compared with the traditional body surface reinforced with bonded prestressed carbon fiber board, the carbon fiber board of this technology is farther from the centroid of the reinforced structure, and the same prestress produces a larger eccentric bending moment, and the reinforcement effect is expected to be more significant. Of course, this technology requires more space, and its appearance is slightly inferior to conventional body surface reinforcement with bonded prestressed carbon plates. However, this technology is suitable for bridge reinforcement under special conditions, such as certain mountain bridges (under-bridge space) . It has good engineering application value under the premise of sufficient clearance for navigation and traffic. Therefore, it is necessary to carry out exploration and research on prestressed carbon fiber board string beam combination reinforcement technology.

Keywords: prestressed carbon fiber board; beam string structure; bridge reinforcement; theoretical analysis

制备参数对泡沫沥青稳定碎石性能的影响

王　亚[1]　王显光[2]
（1. 中国建筑第八工程局有限公司 设计管理总院，上海，201206；
2. 交通运输部 科学研究院，北京，100029）

摘　要：从制备工艺角度，考虑成型方法、击实功、放置时间等因素对泡沫沥青稳定碎石的性能影响。测试了不同空隙率条件下，混合料的物理性能、劈裂强度和疲劳性能。结果表明：大功率的成型方法能够有效地减小混合料的空隙率，提高密实程度；试件的放置时间越长，集料的空隙率也越大；大马歇尔试件比标准马歇尔试件的劈裂强度更高；试件的养护龄期越长，劈裂强度越高；试件成型碾压次数对疲劳性能有着显著的影响；细型级配有助于提高混合料的疲劳寿命。制备参数对指导泡沫混凝土现场施工有着重要的指导意义。
关键词：路面材料；泡沫沥青稳定碎石；空隙率；制备方法；劈裂强度；疲劳特性

Effect of synthesis parameters on the properties of foamed asphalt stabilized macadam

Wang Ya[1]　*Wang Xianguang*[2]
（1. Institute of Design and Management，China Construction Eighth Engineering Division Co.，Ltd.，Shanghai 201206，China；
2. Research Institute of Sciences，Ministry of Transport，Beijing 100029，China）

Abstract：From the perspective of synthesis technology，the influences of compaction method，compaction power and storage time on the performance of foamed asphalt stabilized macadam were considered. The physical properties，splitting strength and fatigue properties of the mixture at different air voids were tested. The results show that the high power compaction method can effectively reduce the air void of the mixture and improve the degree of compaction. The longer the specimen is stored，the larger the air void of aggregate is. The splitting strength of the large Marshall specimen is higher than that of the standard Marshall specimen. The longer the curing age of the specimen，the higher the splitting strength. The number of rolling has a significant effect on the fatigue properties of the specimen. The fine grading is helpful to improve the fatigue life of the mixture. The synthesis parameters are of great significance to the field construction of foamed concrete.
Keywords：pavement materials；foamed asphalt stabilizes macadam；air void；synthesis technology；splitting strength；fatigue performance

偏心受压 GFRP 筋混凝土柱的承载力计算方法

肖 刚 谭 伟 欧进萍
(哈尔滨工业大学，广东 深圳，518055)

摘　要：通过理论分析和试验数据系统性地分析了 GFRP 筋混凝土偏压构件的承载力设计计算方法。首先基于曲率法和 GFRP 筋混凝土柱的偏压试验数据，创新性地构造了考虑相对偏心距、长细比和强度配筋率的偏心距增大系数计算式，并拟合得到了其中的待定参数；然后根据钢筋混凝土柱的原理推导了考虑相对偏心距和长细比的偏心距增大系数简化计算式；最后由平衡方程和变形协调得到了 GFRP 筋混凝土矩形和圆形截面柱的偏压承载力设计计算方法，并通过试验数据进行了验证。结果表明，偏心距增大系数会随相对偏心距递增，且强度配筋率越小或长细比越大，变化越快，偏心距增大系数和偏压承载力的计算值均与试验值吻合良好。

关键词：GFRP 筋；混凝土柱；二阶效应；弯矩放大系数；偏压承载力

Calculation method of ultimate capacity of eccentrically loaded GFRP reinforced concrete columns

Xiao Gang　Tan Wei　Ou Jinping
(Harbin Institute of Technology, Shenzhen, Guangdong 518055, China)

Abstract: Through theoretical analysis and experimental data, the design and calculation method of bearing capacity of eccentrically loaded GFRP reinforced concrete members is systematically analyzed. Based on the curvature method and the eccentric compression test data, the calculation formula of the eccentricity increase factor considering the relative eccentricity, slenderness ratio and strength reinforcement ratio is proposed, and the undetermined parameters are obtained. Then, according to the principle of reinforced concrete columns, the simplified formula of eccentricity increasing factor considering relative eccentricity and slenderness ratio is derived. Finally, based on the equilibrium equation and deformation compatibility, the design and calculation methods of eccentrically loaded GFRP reinforced concrete columns with rectangular and circular cross-sections are obtained and verified by experimental data. The results show that the eccentricity increasing factor increases with the relative eccentricity, and the smaller the strength reinforcement ratio is or the greater the slenderness ratio is, the faster the change is. The calculated values of eccentricity increasing factor and eccentric bearing capacity are in good agreement with the experimental values.

Keywords: GFRP bar; concrete column; second-order effect; moment magnification factor; eccentric bearing capacity

马尔代夫吹填珊瑚礁地基 CFG 桩地基处理设计研究与应用

陈 骏 庞海枫 何 栋 程 伟 叶 伟
（中建三局第一建设工程有限责任公司，湖北 武汉，430040）

摘 要： 马尔代夫 7000 套保障性住房项目位于胡鲁马累围海造地岛屿二期岛，该岛屿属于新近吹砂回填岛屿，本项目首次在当地采用了 CFG 桩和桩端后注浆工艺进行地基处理，因为缺乏吹填珊瑚礁地基处理的设计规范及相关资料，在有代表性的场地上进行现场试验及试验性施工，通过试验确定设计参数和处理效果，指导设计及施工，通过该工程成功应用有效地实现了我国规范及工艺的输出，同时也为类似吹填珊瑚礁地基设计、施工提供了工程参考。
关键词： 吹填珊瑚礁；地基处理；CFG 桩；桩端后注浆；静载试验

Design and application of CFG pile foundation treatment for hydraulic fill coral reef foundation in Maldives

Chen Jun Pang Haifeng He Dong Cheng Wei Ye Wei
(China Construction Third Bureau First Engineering Co.,
Ltd., Wuhan, Hubei 430040, China)

Abstract: Maldives 7000 indemnificatory housing project is located in hulumale reclamation Island phase Ⅱ Island, which is a newly dredged sand backfilled island. For the first time, CFG pile and pile end post grouting technology are used for foundation treatment in this project. Due to the lack of design specifications and relevant data for foundation treatment of dredged coral reef, field test and experimental construction are carried out on representative sites, Through the successful application of the project, the output of Chinese specifications and technology is effectively realized. At the same time, it also provides engineering reference for the design and construction of similar hydraulic fill coral reef foundation.
Keywords: hydraulic fill coral reef; foundation treatment; CFG pile; post grouting at pile end; static load test

三跨上承空腹式飞鸟拱拱圈及板拱施工技术

曾　希　罗思杭　刘愿祝　高　飞　刘锦晖　张驰星
（中建三局第二建设工程有限责任公司安装公司，湖北 武汉，430000）

摘　要：拱桥造型美观大气、景观效果极佳，为城市中常用的桥梁形式之一，其中拱圈和板拱为施工重难点。本文依托咸宁大洲湖项目，对三跨总长约 74.2m 的上承式空腹飞鸟拱进行了施工，支架采用满堂架和少支点贝雷架组合支撑体系，通过支架的设计与搭设、支架预压、钢筋、模板、混凝土工程施工，测点的布设与变形监测、支架的拆除等施工工序，确保了拱桥施工过程中的安全性和较好的成型质量，可为类似市政桥梁工程提供参考。
关键词：拱桥；组合支撑；变形监测

Analysis of the effective implementation for recovering lakes

Zeng Xi　Luo Sihang　Liu Yuanzhu　Gao Fei　Liu Jinhui　Zhang Chixing
(Second Construction Engineering Co., Ltd., China Construction
Third Engineering Bureau, Wuhan, Hubei 430000, China)

Abstract: Arch bridge is one of the commonly used bridge forms in the city with beautiful appearance and excellent landscape effect, in which arch ring and slab arch are the key and difficult points in construction. Based on the project of Xianning Dazhou Lake, this paper has carried out the construction of the three-span hollow bird arch with 74.2m. The support adopts the combined support system of full hall frame and less fulcrum berley frame, through the construction processes such as support design and erection, support preloading, reinforcement, formwork and concrete engineering construction, measurement point layout and deformation monitoring, support removal and so on, the safety and good forming quality in the construction process of arch bridge are ensured, which can provide reference for similar municipal bridge projects.
Keywords: arch bridge; combined support; deformation monitoring

浅谈预应力张拉影响因素

陈佳琪
（中建五局第三建设有限公司，湖南 长沙，410007）

摘　要：在现代桥梁的建设过程中，梁体的预应力张拉存在许多的影响因素。本文主要对实际张拉过程当中，钢绞线的实际伸长值与理论伸长值之间的偏差进行研究，从伸长量计算起点、测试压力表精度以及钢绞线自身力学参数等方面来分析预应力钢绞线伸长量，采取针对性的预控措施，以确保实际伸长值符合规范要求，有效地降低钢绞线实际伸长值与理论计算值之间的偏差，具有一定的工程实际意义。
关键词：预应力张拉；实际伸长值；理论伸长值；影响因素

Talking about the influencing factors of prestress tension

Chen Jiaqi
(China Construction Fifth Bureau Third Construction Co., Ltd.,
Changsha, Hunan 410007, China)

Abstract: In the construction of modern bridges, there are many influencing factors in the prestressing of beams. This article mainly studies the deviation between the actual elongation value of the steel strand and the theoretical elongation value in the actual tensioning process, from the calculation starting point of the elongation, the accuracy of the test pressure gauge, and the mechanical parameters of the steel strand itself. Analyze the elongation of the prestressed steel strand, take targeted pre-control measures to ensure that the actual elongation value meets the requirements of the specification, effectively reduce the deviation between the actual elongation value of the steel strand and the theoretical calculation value, and has a certain The practical significance of the project.
Keywords: prestressed tension; actual elongation value; theoretical elongation value; Influencing factors

基于 Tranformer 模型的安全事故分类的研究

陆梦阳
（中国建筑国际工程有限公司）

摘　要：对以往安全事故的案例分析是进行安全管理的重要环节，世界各国专家学者都在研究事故伤亡的发生规律以寻求解决对策。随着大数据和人工智能的发展，迫切需要一种高效自动对安全事故文本分析的方法，本研究基于自然语言处理的相关技术和方法，运用基于 Transformer 结构的 BERT 模型对安全事故案例进行标签分类，并和基于循环神经网络的模型 Lstm 和 Bilstm 进行了对照。实验证明，BERT 模型分类效果最优，并在测试数据集的评测上能到达平均 90%的 F1 值，优于 Lstm 和 Bilstm 模型。本研究结果表明 Transformer 模型对建筑行业安全事故案例的分类任务表现优异，为安全行业的事故分析提供了一个强大的数字化框架，有利于提高安全行业的工作和生产效率

关键词：安全事故分析；自然语言处理；文本分类；Transformer 模型

Study on classification of safety accidents based on tranformer model

Lu Mengyang
(China State Construction International Engineering Co. , Ltd.)

Abstract: Case analysis of past safety accidents is an important part of safety management. Experts and scholars from all over the world are studying the pattern behind various accidents reports to come up with relative solutions. With the development of big data and artificial intelligence, there is an urgent need for an efficient and automatic method for text analysis of safety accidents. This research is based on natural language processing related technologies and methods, and uses the BERT model based on the transformer structure to classify safety accident cases, in comparison with two other models, Lstm and Bilstm, which are based on recurrent neural network. Experiments have proved that the BERT model has the best performance and can reach an average 90% F1 value in the evaluation of the test data set, which is better than the Lstm and Bilstm models. The results of this study show that the Transformer model performs well in the classification task of safety accident cases in the construction industry, provides a powerful digital framework for accident analysis in the safety industry, and is effective in improving the work and production efficiency of the safety industry.

Keywords: safety accidents analysis; natural language processing; text classification; transformer model

独柱盖梁钢管柱贝雷梁施工技术研究

谷崇建
（中国建筑第五工程局有限公司，湖南 长沙，410004）

摘　要：在桥梁建设中，为优化下部建筑空间、减少河道水阻比等，常常会遇到独柱盖梁的施工。在常用的满堂支架法、穿心棒法、抱箍法难以实施的工况下，独柱盖梁钢管柱贝雷梁施工技术可以有效地解决满堂支架法地基难加固、穿心棒法无法形成有效支撑、抱箍法难以对方形柱有效紧固的问题，同时利用承台做基础解决了地基承载力、堆载预压的问题，减少了工作量和施工周期。本技术成本低、效率高、安全可靠、技术先进，有明显的社会和经济效益。

关键词：独柱；盖梁；钢管柱；贝雷梁

Research on construction technology of bailey beam with single column cap beam and steel tube column

Gu ChongJian
（China Construction No. Fifth Engineering Division Co.，Ltd.，
Changsha，Hunan 410004，China）

Abstract：In bridge construction，in order to optimize the lower building space and reduce the water resistance ratio of the river，the construction of single column bent cap is often encountered. Under the condition that the commonly used full support method，through bar method and hoop method are difficult to implement，the construction technology of single column bent cap steel pipe column Bailey beam can effectively solve the problems that the full support method is difficult to strengthen，the through bar method is unable to form effective support，and the hoop method is difficult to effectively fasten the square column. At the same time，the bearing capacity and surcharge preloading of the foundation are solved by using the pile cap as the foundation，the workload and construction period are reduced. The technology has the advantages of low cost，high efficiency，safety and reliability，advanced technology，and obvious social and economic benefits.

Keywords：single column；bent cap；steel pipe column；bailey beam

强震珊瑚礁砂地区陆上及水下碎石桩抗液化技术研究

胡利文[1]　梁小丛[1]　徐　雄[2]　朱明星[1]　井　阳[2]

（1. 中交四航工程研究院有限公司，广东 广州，510230；

2. 中交四航局第二工程有限公司，广东 广州，510230）

摘　要： 对位于超强地震带上的松散珊瑚礁砂地基，采用基于密实法的抗液化处理技术往往需要满足较高的密实度，而对于含较高细粒（平均细粒含量约22%）原状珊瑚礁砂土，传统振冲密实法较难满足，因此需要采用提供排水为主的碎石桩抗液化处理技术；为了充分利用桩间土的抗液化强度，对碎石桩成桩质量提出了需满足桩间土密实度以及成桩排水通道质量的控制指标。结合东帝汶Tibar港项目，现场选取了不同细粒含量的典型珊瑚礁砂地层，开展了陆域底部出料振冲法及水下底部出料振冲法碎石桩的试验区研究。由试验区结果分析表明，提出的振冲工艺控制参数和质量控制流程能满足排水为主挤密为辅碎石桩质量要求，同时分别对陆域振冲法在不同细粒含量地层不同置换率的试验数据进行对比分析，获取了陆域振冲碎石桩挤密规律。另外基于海域振冲法施工的特殊性，研究分析了海域振冲碎石桩施工工艺，并与陆域振冲挤密效果进行了对比分析。所获取研究成果可为后续类似珊瑚礁砂地基陆域和海域的碎石桩振冲碎石桩地基处理提供参考借鉴。

关键词： 珊瑚礁砂地基处理；碎石桩；抗液化；施工技术

Research on construction technology of（off shore）stone column in coral sand stratum located in strong earthquake zone

Hu Liwen[1]　*Liang Xiaocong*[1]　*Xu Xiong*[2]　*Zhu Mingxing*[1]　*Jing Yang*[2]

（1. CCCC Fourth Harbor Engineering Institute Co.，Ltd.，Guangzhou，Guangdong 510230，China；

2. Second Engineering Company of CCCC Fourth Harbor Engineering Institute Co.，Ltd.，Guangzhou，Guangdong 510230，China）

Abstract：For loose coral sand stratum located in strong earthquake magnitude zone，a higher compaction degree is normally required if vibro-compaction ground improvement is adopted. However，it would not be compacted properly if a higher fines content encounters，i. e. average 22%，in coral sand stratum. Therefore，stone column which could act as drainage function criteria and densification function criteria during earthquake vibration is applied for ground improvement in Tibar Project in Timor-Leste. In order to verify the drainage function and densification function of stone column being achieved in coral sand subsoil，trail tests with different replacement

ratios have been conducted onshore and offshore for bottom feed vibro-flotation separately. Based on the comparative analysis of test results, the quality control parameters and procedure have been proposed to ensure the drainage function and densification of stone column. Further, the densification for different replacement ratio and different subsoils under coral sand onshore have be obtained. Moreover, in consideration of the characteristic of vibro-flotation method performed offshore, the densification and quality control measures are analyzed and summarized comparing with that obtained onshore. The obtained research result could serve as a referred guideline for similar geological condition in coral reef stratum.
Keywords: coral sand ground improvement; stone column; liquefaction mitigation; construction technology

优化神经网络在基坑多因素变形的预测分析

岳建伟[1,2]　仲豪磊[1]　顾丽华[1,2]　邢旋旋[1]　黎　鹏[1,2]　张　静[1,2]　王自法[1]

（1. 河南大学 土木建筑学院，河南 开封，475004；

2. 河南省轨道交通智能建造工程中心，河南 开封，475004）

摘　要：针对传统神经网络用于深基坑施工开挖预测时，网络结构过于复杂而易出现过拟合现象，本文提出遗传算法和贝叶斯正则化组合优化 BP 神经网络模型。首先利用多元线性回归模型对基坑客观因素和人为影响因子进行筛选分析，然后通过贝叶斯正则化对网络权值的限制来简化网络结构，从而有效地抑制过拟合，提高 BP 神经网络全局优化能力以及泛化能力，仿真结果表明：该方法对基坑监测地表沉降和水平位移预测的平均相对误差分别为 2.068%和 4.846%，并首次验证该组合模型应用在深基坑变形预测方向的可行性，为基坑变形预测提供了新的思路和方法。

关键词：贝叶斯正则化；BP 神经网络；遗传算法；深基坑工程；多元线性回归；主客观影响因素

Prediction and analysis of multi-factor deformation of foundation pit by optimized neural network

Yue Jianwei[1,2]　*Zhong Haolei*[1]　*Gu Lihua*[1,2] *Xing Xuanxuan*[1]

Li Peng[1,2]　*Zhang Jing*[1,2] *Wang Zifa*[1]

（1. School of Civil Engineering and Architecture，Henan

University，Kaifeng，Henan 475004，China；

2. Henan Rail Transit Intelligent Construction Engineering Center，

Kaifeng，Henan 475004，China）

Abstract：Aiming at the fact that the traditional neural network is used to predict deep foundation pit construction and excavation，the network structure is too complex and prone to overfitting. This paper proposes a combination of genetic algorithm and Bayesian regularization to optimize BP neural network model. First，use the multiple linear regression model to screen and analyze the objective factors and man-made influencing factors of the foundation pit，and then simplify the network structure through Bayesian regularization to limit the network weights，thereby effectively inhibiting overfitting and improving the overall BP neural network Optimization ability and generalization ability. The simulation results show that the average relative error of this method for the prediction of ground settlement and horizontal displacement of foundation pit monitoring is

2.068% and 4.846%, respectively. It is the first time to verify that the combined model is applied to the prediction direction of deep foundation pit deformation. Feasibility provides new ideas and methods for the prediction of foundation pit deformation.
Keywords: bayesian regularization; BP neural networks; genetic algorithm; deep foundation pit project; multiple linear regression; subjective and objective factors

N-JET 工法桩施工对基坑及环境变形的影响研究

贾文强
（上海市基础工程集团有限公司，上海，200000）

摘　要：上海轨道交通14号线陆家嘴车站主体围护为52m地下连续墙，地墙采用十字钢板接头，基坑开挖深度27.9～29m，开挖难度与风险大，质量要求高，周边环境复杂。为确保基坑开挖安全，实际施工过程中采用N-JET隔水帷幕。结果表明，N-JET隔水帷幕施工对周边环境以及基坑围护均存在一定影响，高压喷浆施工会导致周边地面及构筑物发生抬升，并导致基坑围护产生变形。

关键词：深基坑；N-JET；围护变形；变形控制；构筑物沉降

Influence of N-JET pile construction on foundation pit deformation

Jia Wenqiang
(Shanghai Foundation Engineering Group Co., Ltd., Shanghai 200000, China)

Abstract: The main body of Lujiazui Station on Shanghai Rail Transit Line 14 is enclosed with a 52m underground continuous wall with cross steel plate joint. The excavation depth of the foundation pit is 27.9-29m. It is difficult and risky to excavate with high quality requirements and complicated surrounding environment. In order to ensure the safety of foundation pit excavation, N-JET water-proof curtain is used during actual construction. The results show that the construction of N-JET water-proof curtain has certain influence on surrounding environment and foundation pit retaining structure. High-pressure grouting construction will cause the lifting of surrounding ground and structures and deformation of foundation pit retaining structure .

Keywords: deep foundation pit; N-JET; Deformation of Retaining Structure; Deformation control; Settlement of structures

城市轨道交通工程现浇区间隧道渗漏问题探讨

郑业勇　王维国　赵　峰
（中国建筑第五工程局有限公司，湖南 长沙，410004）

摘　要：城市轨道交通逐渐向复杂网络发展，新线路的并入对缓解城市日益增长的交通压力有着重要的作用。然而，城市轨道交通工程现浇区间隧道仍普遍存在的渗漏水问题对其使用功能及外观质量造成严重影响。文章结合当前渗漏预防与治理工程科技创新探索与实践，分析了区间隧道裂缝渗漏原因及危害，认为渗漏预防与治理是集研究、设计、施工、管理、维护于一体的系统工程，现阶段抗裂防渗成套技术的推广应用，有助于促进城市轨道交通工程高质量发展。

关键词：城市轨道交通；隧道渗漏；抗裂防渗技术；渗漏治理

Discussion on leakage of cast-in-situ tunnels in urban rail transit

Zheng Yeyong, *Wang Weiguo*, *Zhao Feng*
(China Construction Fifth Engineering Division Co., Ltd.,
Changsha, Hunan 410004, China)

Abstract: Urban rail transit is gradually developing into a complex network, and the incorporation of new lines plays an important role in alleviating the growing traffic pressure in the city. Combined with a large number of cases and documents, it can be known that the common leakage problems in cast-in-situ tunnels of urban rail transit projects are closely related to factors such as material performance, design level, construction control and post-maintenance. Based on the current engineering practice of leakage prevention and leakage treatment, the article summarizes the causes and hazards of the tunnel crack leakage, and believes that the prevention and treatment of leakage is a systematic project that integrates research, design, construction, management, and maintenance. The popularization and application of a complete set of anti-cracking and anti-seepage technologies at the stage will help promote the high-quality development of urban rail transit.

Keywords: urban rail transit; tunnel leakage; anti-cracking and anti-seepage technology; leakage treatment

施工间隔时间对公建屋面防水粘结层施工质量的影响研究

吴开放　黄易平　何海涛　李　杨

（中建五局第三建设有限公司，湖南 长沙，410004）

摘　要： 由于公建屋面防水粘结层每层材料的施工间隔时间没有规律，在施工时会造成因工期延误而导致防水粘结层粘结效果不佳的问题，甚至影响到整个屋面铺装的质量，为此研究施工间隔时间对公建屋面防水粘结层的影响十分重要。首先准备试验材料，将与面板材质及厚度相同的钢板制作成试件，采用 Posi Test AT 拉拔强度测试仪进行拉拔试验，试验所得结论为：建议防腐层与甲基丙烯酸甲酯树脂膜施工的间隔时间不超过 4d，聚合物改性沥青与粘结剂的施工间隔时间不超过 6d。最后结合现场检测分析确定了最佳的间隔时间为 1～3d。

关键词： 施工间隔时间；公建屋面；防水粘结层；施工质量；渗透；铺装结构

Research on the influence of construction interval on the construction quality of public building roof waterproof cohesive layer

Wu Kaifang　Huang Yiping　He Haitao　Li Yang

(3rd Construction Co., Ltd. of China Construction 5th Engineering Bureau, Changsha, Hunan 410004, China)

Abstract: Since the construction interval of each layer of the waterproof bonding layer of public buildings is irregular, the construction will cause the problem of poor bonding effect of the waterproof bonding layer due to the delay of the construction period, and even affect the quality of the entire roofing. For this reason, the influence of the construction interval on the waterproof bonding layer of public roof is studied. First, prepare the test materials, make the steel plate of the same material and thickness as the panel into the test piece, and use the Posi Test AT pull-out strength tester to carry out the pull-out test. The conclusion of the experiment is: it is recommended to construct the anti-corrosion layer and methyl methacrylate resin film The interval between the construction of the polymer modified asphalt and the binder does not exceed 4d, and the construction interval between the polymer modified asphalt and the binder does not exceed 6d. Finally, combined with on-site inspection and analysis, the best interval time is determined to be 1-3d.

Keywords: construction interval; public building roof; waterproof bonding layer; construction quality; penetration; pavement structure

不同标准对不同类型地下水及不同条件抗浮设防水位取值对比与分析

彭柏兴

（长沙市规划勘测设计研究院，湖南 长沙，410007）

摘 要： 抗浮设防水位的确定对地下结构的安全使用至关重要，地下结构因抗浮而产生的变形、破坏事件时有发生，抗浮设防问题已成为工程界的热点。地下水位的影响因素甚多，包括大气降水、地质 、水文条件和人为因素。近年来，对抗浮设防水位的研究成果丰富，各级规范、标准中都有涉及，但分歧依然存在，一定程度上影响了工程实际应用。本文综合分析了近年的研究成果，对 24 本国家、行业、地方标准关于抗浮设防水位的规定进行了归纳与梳理，以地下水的类型为主线，结合特殊场地与施工要求，提出了地下水位抗浮设防取值的基本原则，对未来工作提出了展望与建议。

关键词： 地下结构；抗浮设防水位；上层滞水；潜水；承压水；岩溶水；特殊场地

Analysis and discussion on water level prevention of up-floating according of different types of groundwater according & different environmental conditions to different standards

Peng Baixing

(Changsha Urban Planning , Investigation & Design, Insitute, Changsha, Hunan 410007, China)

Abstract: How to determinate the water level prevention of up-floating is very important for the safty of underground structures. The deformation and damage events of underground structures occur frequently on account of up-floating, the problem of anti-floating has become a hot spot in engineering field. The influence factors of groundwater levelinclude rainfalls, soil and hydrogeology conditions and human activity. In recent years, there are abundant research results on the determination of anti-floating water level, which are involved in different specifications and standards, but there are still differences, which affect the practical application for project. The provisions of anti-floating water level are summarized and sorted out from 24 national, industrial and local standards on the base of comprehensive analysis. And then, the basic principle of anti-floating prevention value of groundwater level value are proposed by taking the type of groundwater as the main line, combined with the special site and construction requirements, etc. , and the prospect and suggestions for future work are put forward.

Keywords: underground structures; water level prevention of up-floating; perched water; phreatic water; confined water; special site

抗滑桩与石笼挡墙组合防护技术

钟华湘[1]　刘　泽[2]
（1. 中国建筑第五工程局有限公司，湖南 长沙，410004；
2. 中国建筑股份有限公司阿尔及利亚公司，阿尔及尔，16101）

摘　要：半刚性石笼挡墙结构具有适应沉降变形的特点，变形后仍具有整体性，其施工简单方便，在非洲砂性土地区应用广泛。抗滑桩在国内外滑坡体治理工程中均有应用。无论国内还是国外，抗滑桩、石笼挡墙防护技术单独应用案例较多，但两者组合防护却未见记载。通过对阿尔及利亚南北高速公路项目梅地亚绕城段滑坡治理进行研究，发现大面积滑坡地段采用单排抗滑桩与石笼挡墙相结合的防护技术具有良好效果。石笼挡墙成本相对较低，能够降低工程造价；施工方法简单，容易满足质量要求，还能缩短施工工期。沉降变形观测结果表明，抗滑桩与石笼挡墙相结合的防护技术对于长大滑坡体治理具有显著作用，可以应用于浅层和中厚层滑坡体治理。

关键词：抗滑桩；石笼挡墙；滑坡体；边坡防护

Combined protection technology of anti-slide pile and stone cage retaining wall

Zhong Huaxiang[1]　*Liu Ze*[2]
（1. China Construction Fifth Engineering Division Corp.，
Ltd.，Changsha，Hunan 410004，China；
2. Algeria Branch of China State Construction Engineering Corp.，
Ltd.，Algiers，16101，Algeria）

Abstract：The structure of semi-rigid gabion retaining wall adapts to settlement and deformation，and remains integral after that. Its construction is simple and convenient，and it is widely used in sandy soil areas in Africa. Anti-slide piles are used in landslide treatment projects in China and abroad. However，there are many cases of independent application of anti-slide pile and gabion retaining wall protection technology at home and abroad，but the combined protection of the two has not been recorded. Through the study on the treatment of the landslide in the ring road in Médéa Province of the North-south Highway Project（53 km）in Algeria，it is found that the protection technology combined with single-row anti-slide piles and gabion retaining walls in large-area landslide areas has good results. The cost of the gabion retaining wall is relatively low，which can reduce the project cost；the

construction method is simple, easy to meet the quality requirements and can shorten the construction period. The observation results of settlement and deformation show that the combination of anti-slide piles and gabion retaining walls has a significant effect on the treatment of long and large landslides, and can be applied to the treatment of shallow and medium-thick landslides.
Keywords: anti-slide pile; gabion retaining wall; landslide body; slope protection

智能化粮库建设及应用

姜志浩　刘　玮　王　恒　李柔锋　何　彬
（中建三局集团有限公司，湖北 武汉，430000）

摘　要： 智能化粮库建设是我国粮食仓储企业适应信息化发展的必然趋势，针对智慧化粮库目前发展问题，利用计算机、网络、传感器、自动检测、自动化控制和视频监控监测等技术，搭建智能化粮库管控集成控制平台，基于平台研发智能粮食仓储、智能粮食出入库、智能粮库安防等系统，实现对库区粮食安全储藏的现代化管理。
关键词： 智能化；集成控制平台；智能仓储；智能出入库

Construction and application of intelligent grain depot

Jiang Zhihao　Liu Wei　Wang Heng　Li Roufeng　He Bin
（China Construction Third Engineering Bureau Co.，Ltd.，
Wuhan，Hubei 430000，China）

Abstract：The construction of intelligent grain depots is an inevitable trend for our country's grain storage enterprises to adapt to the development of informatization. Aiming at the current development problems of intelligent grain depots，the use of computers，networks，sensors，automatic detection，automatic control and video monitoring and other technologies to build intelligent grain The warehouse management and control integrated control platform，based on the development of intelligent grain storage，intelligent grain in and out of storage，intelligent grain storage security and other systems，realizes the modern management of the safe storage of grain in the reservoir area.
Keywords：intelligent；integrated control platform；intelligent warehousing；intelligent warehousing

基于钢盾构箱涵斜交顶进力计算程序开发与应用

辛亚兵[1]　陈　浩[2]　谭　鹏[1]　胡富贵[1]
（1. 湖南建工交通建设有限公司，湖南 长沙，410004；
2. 湖南建工集团有限公司，湖南 长沙，410004）

摘　要：为正确、方便地计算斜交地道桥最大顶进力，以湖南平江至益阳高速公路 NK1＋100.2 箱涵顶进施工为工程背景，首先根据力学平衡原理推导了地道桥顶进力最大值计算公式，然后利用计算机 C# 语言编制了基于钢盾构施工地道桥顶进力计算软件；在此基础上，分别对 NK1＋100.2 地道桥顶进力最大值按手算和编程软件方法进行计算，并进行了对比分析。结果表明，大交角地道桥顶进施工两侧顶进力最大值相差较大；基于钢盾构施工地道桥顶进力计算软件计算结果可靠，具有操作简单，方便实用的优点。

关键词：地道桥；最大顶进力；大交角；计算分析；BIM 模型

Development and application of calculation program for oblique jacking force of steel shield box culvert

Xin Yabing[1]　*Chen Hao*[2]　*Tan Peng*[1]　*Hu Fugui*[1]
（1. Hunan Construction Engineering and Transportation Construction Co., Ltd., Changsha, Hunan 410004, China;
2. Hunan Construction Engineering Group Co., Ltd., Changsha, Hunan 410004, China）

Abstract: In order to calculate the maximum jacking force of skew underpass bridge correctly and conveniently, taking the jacking construction of NK1＋100.2 box culvert of Pingjiang Yiyang Expressway in Hunan Province as the engineering background, the calculation formula of the maximum jacking force of underpass bridge was deduced according to the principle of mechanical balance, and then the calculation program of jacking force of underpass bridge based on steel shield construction was compiled by the computer C# language. On this basis, the maximum jacking forces of NK1＋100.2 underpass bridge were calculated by hand and the programming software method, and the comparative analysis was carried out. The results show that there is a significant difference in the maximum jacking force between the two sides of the tunnel bridge with large intersection angle. The calculation results of the jacking force calculation program based on steel shield construction tunnel bridge are reliable, easy to operate, convenient and practical.

Keywords: underpass bridge; maximum jacking force; large angle of intersection; calculation and analysis; BIM model

香港建造业环保制度浅析

姜海峰[1]　姚泽恒[2]　李易峰[2]　纪汗青[2]　曲晓阳[2]
（1. 中国海外集团有限公司，中国 香港，999077；
2. 中国建筑工程（香港）有限公司，中国 香港，999077）

摘　要： 合理的环保管理组织结构、健全的环境法律制度、有效的实施和监督体系构成了香港建造业先进的环保制度。以中建香港承建的搬迁沙田污水处理厂往岩洞地盘为参考，分析香港建造业环保制度的发展与构成，结合地盘环保管理实践，总结以环境影响评估制度和多方参与为核心的香港环保制度实施与监督体系，为内地工程环保制度的发展完善提供参考和建议。

关键词： 环保制度；香港建造业；环境影响评估；工程监督

Study on environmental protection system of Hong Kong construction industry

Jiang Haifeng[1]　*Yao Zeheng*[2]　*Li Yifeng*[2]　*Ji Hanqing*[2]　*Qu Xiaoyang*[2]
(1. China Overseas Holdings Ltd., Hong Kong, China;
2. China State Construction Engineering (Hong Kong) Ltd., Hong Kong 999077, China)

Abstract: Reasonable management and organization structures, a comprehensive legal system, and an effective implementation and supervision system constitute the advanced environmental protection system of Hong Kong construction industry. Based on 'Relocation of Sha Tian Sewage Treatment Works to Caverns-Site Preparation and Access Tunnel Construction' contracted by China State Construction Engineering (Hong Kong) Ltd., the development and composition of the environmental protection system of Hong Kong construction industry were analyzed. Combined with site practice, the implementation and supervision system featured by environmental impact assessment and multi-party participation was summarized. Valuable references and suggestions are generated for the development of environmental protection system for construction projects in mainland China.

Keywords: environmental protection system; Hong Kong construction industry; environmental impact assessment; engineering supervision

轨道非接触式无损检测技术数值模拟研究

周　涛　葛　浩
（长江勘测规划设计研究有限责任公司，湖北 武汉，430010）

摘　要：考虑到现有钢轨检测技术的不足，提出了基于空气耦合导波的钢轨非接触式无损检测方法，建立了可模拟空气耦合导波激励与接收全过程的声固耦合仿真模型，并基于声学理论对仿真模型进行了验证。通过该数值模型模拟分析了轨底不同损伤程度对接收导波信号的影响。结果表明，钢轨空气耦合接收导波仍具有能量集中的特点，不同损伤程度对应损伤指数范围有所差异，基于空气耦合导波检测技术可对不同钢轨损伤进行定量化评估。

关键词：钢轨；导波；数值模拟；损伤评估

Research on numerical simulation of rail non-contact nondestructive testing technology

Zhou Tao　Ge Hao
(Changjiang Institute of Survey, Planning, Design and Research, Wuhan, Hubei 430010, China)

Abstract: Considering the shortcomings of the existing rail inspection technology, a non-contact nondestructive inspection method based on air coupled guided wave is proposed. An acoustic-solid coupling simulation model is established to simulate the whole process of air coupled guided wave excitation and reception, and the simulation model is verified based on acoustic theory. Through the numerical model, the influence of different damage severity of rail bottom on the received guided wave signal is simulated and analyzed. The results show that the air coupled guided wave still has the advantage of energy concentration, and the range of damage index will be different with the change of damage severity. Based on the air coupled guided wave detection technology, different rail damage can be quantitatively evaluated.

Keywords: rail; guided wave; numerical simulation; damage assessment

高层建筑深基坑内撑式排桩支护结构变形模拟研究

覃安松　冉光强　蹇林君　潘小勇　严致远　吴楚湘
（中建五局第三建设有限公司，湖南 长沙，410000）

摘　要：内撑式排桩支护结构具有抑制侧向变形、侧向刚度大等优势，在高层建筑深基坑围护中的应用较广，但现阶段对于深基坑变形模拟的研究不够深入，因此本文以某高层建筑施工深基坑工程为例，根据工程概况以及支护方案，建立有限元三维分析模型。通过计算等效刚度以及土压力等，以实现高层建筑深基坑内撑式排桩支护结构变形模拟。试验结果表明，模拟得到的不同工况下水平位移随深度的变化规律与标准结果大致相同，且计算结果与标准结果几乎一致，可较好反映基坑实际开挖情况；当桩长较小时会导致被动土压力不足，持续增加桩长会使桩长对基坑周围土体位移的影响下降；基坑周围土体水平位移受到超载的影响较大，当超载为45kPa时，位移最大值达到22mm；支撑温度升高对第三道钢支撑的横、纵、角支撑影响最大，支撑温度持续提升会使钢支撑失去稳定性，温度降低过大会导致支护能力丢失。

关键词：高层建筑；深基坑；内撑式排桩；支护结构；有限元；土压力

Simulation study on deformation of inner bracing row pile supporting structure in deep foundation pit of high rise building

Qin Ansong　Ran Guangqiang　Jian Linjun
Pan Xiaoyong　Yan Zhiyuan　Wu Chuxiang
(The Third Construction Co., Ltd. of China Construction Fifth Engineering Bureau, Changsha, Hunan 410000, China)

Abstract: With the advantages of restraining lateral deformation and large lateral stiffness, the internal bracing row pile support structure is widely used in the deep foundation pit support of high-rise buildings. However, the research on the deformation simulation of deep foundation pit is not deep enough at this stage. Therefore, this paper takes a deep foundation pit project of a high-rise building construction as an example, establishes a finite element three-dimensional analysis model according to the project overview and support scheme, and calculates the equivalent value. In order to realize the deformation simulation of high-rise building deep foundation pit inner bracing row pile supporting structure, the stiffness and earth pressure are analyzed. The experimental results show that the variation rule of horizontal displacement with depth is similar to the standard results under dif-

ferent working conditions, and the calculation results are almost the same with the standard results, which can better reflect the actual excavation of foundation pit; when the pile length is small, the pressure of the driven soil is insufficient, and the influence of pile length on the displacement of soil around the foundation pit will be decreased by increasing the pile length continuously; the soil around the foundation pit will decrease Horizontal displacement is greatly affected by overload. When the overload is 45kPa, the maximum displacement value reaches 22mm; the rise of support temperature has the greatest influence on the transverse, longitudinal and angular support of the third steel support, and the continuous rise of the supporting temperature will cause the steel support to lose stability, and the excessive temperature reduction will result in the loss of support capacity.
Keywords: high-rise building; deep foundation pit; row of internally supported piles; supporting structure; finite element; earth pressure

杭州萧山国际机场三期新建航站楼“荷花谷”深化设计关键技术

徐丽萍　何　伟　沈小达

（浙江东南网架股份有限公司，浙江 杭州，311209）

摘　要：杭州萧山国际机场三期是浙江省大通道建设十大标志性项目，也是2022年杭州亚运会重要基础配套项目。新建航站楼外形开阔流畅，内部40组变截面钢管柱形如细高的“荷叶”与穹顶钢桁架相连向周围延伸，1组管桁架结构构造的盛开“荷花”立于东侧与交通中心连接区域，整体效果充满了优美感和科技感。本文重点阐述了这1组盛开“荷花”（以下简称“荷花谷”）在深化设计过程中的重难点及关键技术。“荷花谷”有2个组成部分，即荷花谷柱、荷花谷柱顶天窗桁架，前者为管桁架结构，后者为箱形桁架结构。整个花瓣结构造型复杂、节点深化加工难度大、高空作业多、焊接质量高。在深化设计中，针对这些重难点，通过优化相关连接节点、开发定位深化技术等方式，最终绘制出了满足建筑结构要求、适用于加工和安装的施工详图。所阐述的深化关键技术，实际结果表明安全可靠、有效快捷，保障了工程的顺利进行。

关键词：桁架；铸钢件；空间弯扭圆管；定位

The key technology of detailed design about "Lotus bud " in New Terminal of Hangzhou International Airport Phase Ⅲ

Xu Liping　He Wei　Shen Xiaoda

(Zhejiang Southeast Space Frame Co. , Ltd. , Hangzhou, Zhejiang 311209, China)

Abstract: Hangzhou International Airport Phase Ⅲ is one of the ten symbolic projects in Zhejiang Province, which is also an important foundation for the 2022 Asian Games in Hangzhou。The new terminal has an open and smooth appearance, with 40 groups of variable section steel tubes in the interior. This structural modeling likes the leaf of lotus . A group of pipe truss structure stands on the east side of the connecting area with the traffic center. The overall effect is full of beautiful and technical sense. This paper focuses on the difficulties and key techniques of this structure (hereinafter referred to as " lotus bud") in the process of detailed design. " Lotus bud" is composed of two parts, namely lotus column and skylight truss. The former is a pipe truss structure, and the latter is a box truss structure. The whole petal structure is complicated, the nodes are difficult to detailed de-

sign and process, and the welding quality is high. Aiming at these difficult points, the detailed design of this structure use new techniques , like optimizing the nodes and developing technology of position . The actual results show these key technology is safe and effective , which ensure the success of the project.
Keywords: truss; steel casting; space bending steel pipe; position

基于有限元的张弦梁高钒索张拉施工模拟分析

刘 谢 刘圣国 蒋 卫 尹 强 戴超虎

（中建五局第三建设有限公司，湖南 长沙，410116）

摘 要：威海国际经贸交流中心多功能展厅屋盖为大跨度张弦梁结构体系且采用双索设计，预应力钢索张拉施工前结构刚度较弱，预应力索张拉过程会形成力二次分配，预应力索张拉顺序对结构变形、安装质量有非常大的影响。高钒索张拉施工前对张拉过程进行施工模拟计算，根据模拟计算和现场施工监测结果，指导张弦梁吊装、张拉、施工设备及支撑措施的拆除工作。解决了张弦梁结构体系在预应力钢索张拉施工前结构刚度较弱，预应力索张拉施工结构变形的难题，保证了安装质量，满足现场施工要求。

关键词：张弦梁；高钒索；有限元分析；张拉；施工模拟分析

Abstract: The roof of the multifunctional exhibition hall of Weihai International Economic and trade exchange center is a long-span tension string beam structure system and adopts double cable design. The structural stiffness is weak before the prestressed steel cable tensioning construction, and the secondary distribution of force will be formed during the prestressed cable tensioning process. The tensioning sequence of prestressed cable has a great impact on the structural deformation and installation quality. Before the construction of high vanadium cable tensioning, the construction simulation calculation shall be carried out for the tensioning process. According to the simulation calculation and on-site construction monitoring results, the lifting and tensioning of beam string, the demolition of construction equipment and support measures shall be guided. It solves the problem of weak structural stiffness and structural deformation of beam string structure system before prestressed steel cable tensioning construction, ensures the installation quality and meets the on-site construction requirements.

Keywords: beam string; high vanadium cable; finite element analysis; tensioning; construction simulation analysis

基于集对分析法的高等级公路 PPP 项目社会资本方关键风险分析

陈镜丞　林剑锋

（中建五局第三建设有限公司，湖南 长沙，410116）

摘　要：为了科学合理地对高等级公路 PPP（Pubilc-Privater Partnership）项目风险管理中社会资本方的关键风险进行判断，以减小项目风险管理的难度。本文通过实际案例和文献资料分析，并通过问卷调查打分的方式，得出了包括政治、经济、市场、建设、运营及不可抗力六类关键风险指标的高等级公路 PPP 项目社会资本方关键风险测度指标体系。结合经验与粗糙集理论，确定了社会资本方关键性风险测度指标的综合权重，并基于集对分析法，建立了高等级公路 PPP 项目社会资本方关键风险评价模型。最后，以湖南某实际工程为案例进行分析，结果表明该模型可有效判断高等级公路 PPP 项目社会资本方的关键风险等级，为高等级公路 PPP 项目社会资本方进行关键风险管理提供了更可靠的依据。

关键词：高等级公路 PPP 项目；风险管理；社会资本；粗糙集理论；集对分析法

Abstract: In order to reduce the difficulty of risk management for high-grade road PPP (Pubilc-Privater Partnership) project, we need to propose a scientific and reasonable method to judge the key risks of the social capital side. Through case analysis, literature research and questionnaire survey, this article has established a key risk measurement index system for the social capital side of high-grade road PPP projects, which contains six key risk indicators such as Politics, Economy, Market, Construction, Operation and Force majeure. The comprehensive weight of key risk measurement indicators of social capital is determined through the combination of engineering experience and Rough Set Theory, and the key risk assessment model of social capital in high-grade road PPP projects is established based on Set Pair Analysis. Finally, this paper makes a case study of a actual project in Hunan. Results shows that this model can help social capital to judge the key risk level more effectively in high-grade road PPP project, and provides a reliable basis for the risk management.

Keywords: high-grade road PPP project; risk management; social capital; rough set theory; set pair analysis

230m 超高层建筑结构选型及构件敏感性研究

金　天　姜　江
（中海企业集团上海公司，上海，200092）

摘　要：超高层建筑需要巨大投资，且资金回报期长。对超高层建筑而言，结构造价比例可达20％～30％。因此，降低结构造价、提高结构性价比尤为重要。本文首先对实际230m超高层建筑结构做了合理选型，并依据上海市装配式政策要求给出了预制装配式整体设计方案。之后，采用等增量敏感性分析方法对该超高层结构进行了优化，结果表明，该方法能高效地将材料分配到各个构件中，有效减小超高层建筑结构的层间位移角，节约结构材料造价。
关键词：超高层建筑；结构选型；结构优化；敏感性分析；层间位移角

Selection of 230m super high-rise structure and research on component sensitivity

Jin Tian　Jiang Jiang
（China Shipping Enterprise Group Co.，Ltd.，Shanghai 200092，China）

Abstract：Super high-rise building needs huge investment and long return period. For super high-rise buildings，the proportion of structural cost can reach 20％-30％. Therefore，it is particularly important to reduce the structural cost and improve the structural cost performance. Firstly，the paper makes a reasonable selection of the 230m super high-rise building structure，and gives the overall design scheme of prefabricated structure according to the policy requirements of Shanghai. Then，the equivalent incremental sensitivity analysis method is used to optimize the super high-rise structure. The results show that the method can efficiently distribute the materials to each component，effectively reduce the story drift angle of the super high-rise building structure，and save the structure cost of material.
Keywords：super high-rise building；structure selection；structural optimization；sensitivity analysis；story drift

岩溶发达区域及软土地基组合作用下锤击沉桩施工质量控制

卓玉霞　陈　凯　齐　晓　张志峰　刘毅力
（中国建筑第五工程局有限公司，湖南 长沙，410000）

摘　要：针对红光物流园项目桩基础施工遇到的复杂岩溶、深厚软土地基等地质情况，通过试桩，得出最合理相关施工技术参数，为后续预应力管桩展开施工提供理论依据，对类似地质条件预应力管桩施工具有一定的借鉴意义。
关键词：岩溶；深厚软土地基；锤击沉桩

Construction quality control of hammer driven pile under the combined action of karst developed area and soft soil foundation

Zhuo Yuxia　Chen Kai　Qi Xiao　Zhang Zhifeng　Liu Yili
(China Construction Fifth Engineering Division Co., Ltd.,
Changsha, Hunan 410000, China)

Abstract: In view of the geological conditions such as complex karst and deep soft soil foundation encountered in the construction of model pile foundation of Hongguang logistics park, the most reasonable relevant construction technical parameters are obtained through pile test, which provides a theoretical basis for the subsequent construction of prestressed pipe pile, and has a certain reference significance for the construction of prestressed pipe pile under similar geological conditions.
Keywords: karst; deep soft soil foundation; hammer driven pile

超重型钢构件新型吊装施工技术

周大伟　贾　岩

（中国建筑第五工程局有限公司，辽宁 沈阳，110000）

摘　要：型钢混凝土结构是在型钢结构的外面包裹有一层钢筋混凝土的外壳，不受含钢率限制，刚度，承载力高，延展性高，随着科学技术的不断进步，以及我国经济的发展，这种结构现已广泛应用于高层建筑和大跨度建筑工程转换层中。而超重型钢构件新型吊装施工技术是目前型钢构件安装施工效率、安装精度、安装安全性较高的一项工法，该施工工法可在作业面狭小的情况下合理组织施工，优化型钢构件加工方式，较少现场焊接和大型机械设备的使用，提高型钢构件的施工效率，从而显著提高工程施工进度、降低施工成本。

关键词：吊装；型钢构件；超重型；施工技术

Super heavy steel members new hoisting construction technology

Zhou Dawei　*Jia Yan*

(China Construction Fifth Engineering Co. , Ltd. , Shenyang, Liaoning 110000, China)

Abstract: Steel reinforced concrete structure is in the shape of steel structure coated with a reinforced concrete shell, with the continuous progress of science and technology and the development of economy in China, this kind of structure has been widely used in high-rise building and long-span building engineering transfer floor. The new hoisting construction method for super-heavy steel members is a construction method with high installation efficiency, installation precision and installation safety at present. This construction method can reasonably organize construction under the condition of narrow working surface, optimizing the processing mode of section steel members, reducing the use of field welding and large-scale mechanical equipment, improving the construction efficiency of section steel members, thus significantly improving the construction schedule and reducing the construction cost.

Keywords: hoisting; section steel member; super heavy; construction technology

磷石膏-二灰系道路基层复合胶凝材料研究

刘春舵[1] 孔德文[2] 宁朝阳[1] 黄均华[1] 杨超超[1]

（1. 中海建筑有限公司，贵州 贵阳，550081；2. 贵州大学，贵州 贵阳，550025）

摘 要： 二灰（石灰与粉煤灰）基路面基层材料具有收缩大，强度低等缺陷，限制了其在工程中更广泛的运用。为解决此关键问题，且大量消耗工业固废磷石膏和粉煤灰，研究将磷石膏和水泥掺入二灰中，通过改变材料组分的相对掺量来调控磷石膏-二灰系道路基层复合胶凝材料的物理及力学性能，并分析其作用机理。结果表明，磷石膏：水泥相对比例的增大能够延缓浆体的水化速率，致使抗压强度显著降低，但能够显著降低基体的孔隙率，有利于改善其耐水性能。而硬化体的收缩率随着磷石膏的增大和养护龄期的延长而不断减小，掺有磷石膏：水泥=53：4的试样养护至28d膨胀率为1.077%，具有微膨胀性。石灰相对掺量的适量增加则能够提供碱性环境以加快生成水化产物，致密基体内部结构以改善其耐水性能，而基体的力学性能，特别是早期抗压强度在石灰：粉煤灰=1：2～1：4时能得到显著的增强。

关键词： 磷石膏；二灰；基层材料；性能调控；微观机理

Resesrch on phosphogypsum modification lime fly-ash road base composites

Liu Chunduo[1] *Kong Dewen*[2] *Ning Chaoyang*[1] *Huang Junhua*[1] *Yang Chaochao*[1]

(1. Guizhou branch of China Shipping Construction Co., Ltd., Guiyang, Guizhou 550081, China;

2. Guizhou University, Guiyang, Guizhou 550025, China)

Abstract: Lime-fly ash-based pavement base material has the defects of large shrinkage and low strength, which limits its wider application in engineering. In order to solve this key problem and consume a large amount of industrial solid waste phosphogypsum and fly ash, the physical and mechanical properties of phosphogypsum-lime-fly ash composite cementitious materials for road base were studied by mixing phosphogypsum and cement into lime-fly ash, and the mechanism of action was analyzed. The results show that the increase of the ratio of phosphogypsum to cement can delay the hydration rate of the paste, resulting in a significant decrease in compressive strength, but can significantly reduce the porosity of the matrix, which is beneficial to improve its water resistance. However, the shrinkage of hardened samples decreased with the increase of phosphogypsum and curing age, and the swelling rate of samples mixed with phosphogypsum: cement = 53 : 4 was 1.077% after

curing for 28 days, which showed slight swelling. Appropriate increase of lime content can provide alkaline environment to accelerate the formation of hydration products and compact the internal structure of the matrix to improve its water resistance, while the mechanical properties of the matrix, especially the early compressive strength, can be significantly enhanced when the ratio of lime to fly ash is 1 : 2-1 : 4.
Keywords: phosphogypsum; lime fly-ash; base material; performance regulation; microscopic mechanism

论电子档案在高速公路建设中的应用

黄均华[1,2]　刘春舵[1,2]　宁朝阳[1,2]　赵长龙[1,2]　胥琳琳[1,2]
（1. 中海建筑有限公司；2. 贵州雷榕高速公路投资管理有限公司）

摘　要： 1988年10月31日，历经四年时间，沪嘉高速公路建成通车，标志着我国内地第一条高速公路的诞生，从此，世界第一高的高速公路桥、世界上最长的沙漠高速公路和世界上最长跨海大桥都在我国建成并通车。截至2020年，全国高速公路通车总里程达到16万km。在高速公路飞速发展的今天，高速公路建设规模大、工期紧、内业任务重；传统档案资料管理模式主要为人工编制、存档，存在各项内业资料编制不规范、查询检索不便、保管难度高、人力资源投入大等困难，已经难以满足当下发展需求。针对这一现状，立足于《中华人民共和国档案法》（简称《档案法》），提出内业资料电子档案管理系统，使用合法的电子签章，以电子档案“单轨制”运行管理，减少人力资源成本、提高内业资料编制效率，同时确保内业资料真实性、完整性、可用性和安全性。

关键词： 高速公路；电子档案；系统；电子签章

Implement electronic archives management to promote the development of expressway informatization

Huang Junhua[1,2]　*Liu Chunduo*[1,2]　*Ning Chaoyang*[1,2]
Zhao Changlong[1,2]　*Xu Linlin*[1,2]
（1. China Construction International Investment (Guizhou) Co., Ltd.;
2. Guizhou leirong Expressway Investment Management Co., Ltd.）

Abstract: On October 31, 1988, after four years, the Shanghai-Jiamen Expressway was completed and opened to traffic, marking the birth of the first expressway in mainland China. Since then, the world's tallest expressway bridge and the world's longest desert expressway as well as the longest cross-sea bridge in the world has been completed and opened to traffic in China. As of 2020, the total mileage of highways in the country will reach 160,000 kilometers. With the rapid development of expressways today, expressway construction is of large scale, tight construction period, and heavy internal tasks; the traditional file management mode is mainly manual compilation and archiving, and there are various of internal business data compilations that are not standardized, inconvenient to inquire and retrieve, and difficult to keep. Difficulties such as high costs and large investment in human resources have made it difficult to meet the current development needs. In response to

this situation, based on the " Archives Law of the People's Republic of China" referred to as " Archives Law", an electronic file management system for internal business data was proposed, using legal electronic signatures, and operating as a " monorail system" for electronic files. Management, reduce human resource costs, improve the efficiency of compilation of internal business data, while ensuring the authenticity, integrity, availability and security of internal business data.
Keywords: highway; electronic file; system; electronic signature

NDB 在工程筹划及结构施工应用阶段的应用实践

李吉顺　陈　楠
（上海建工二建集团有限公司，上海，200080）

摘　要：近年来，传统的建筑运维模式越来越难以满足实际的建设要求，故提出了在工程筹划阶段及机构施工阶段应用 NDB，并实际应用于新开发银行总部大楼，该项目从工作筹划开始就应用 NDB。本工程主要是以 BIM 模型为应用体系的基础，介绍了新开发银行总部大楼 NDB 在项目建设过程中实际的运维效果。
关键词：新开发银行；BIM 技术；模型

Application of NDB in engineering planning and structure construction

Li Jishun　Chen Nan
(Shanghai Construction No. 2 (Group) Co., Ltd., Shanghai 200080, China)

Abstract: In recent years, the traditional operation and maintenance mode of building is more and more difficult to meet the actual construction requirements, so it is proposed to apply NDB in the project planning stage and the organization construction stage. And it is applied to the headquarters building of the new development bank. NDB has been applied in the project since the work planning. This project mainly based on BIM model, introduces the actual operation and maintenance effect of NDB in the construction process of the New Development Bank headquarters building.
Keywords: New Development Bank; BIM Technology; model

逆作法“一柱三桩”托换体系竖向荷载传递规律研究

邓　亮[1,2]

（1. 上海建工二建集团有限公司，上海，200080；
2. 上海建筑工程逆作法工程技术研究中心，上海，200080）

摘　要：为研究逆作法“一柱三桩”托换体系的力学响应规律，采用自主研发的自动化监测系统在虹桥进口商品展示交易中心（二期）4#楼的施工建造过程中对“一柱三桩”体系的竖向承力结构进行实时监测。监测结果表明：1）上下同步逆作法施工过程中，B0 板上的“一柱三桩”托换体系竖向结构存在周期性的“拉压转换”，且与开挖工况存在因果关系；2）在“一柱三桩”托换体系施工过程中，B0 钢柱承担了大约 50%的竖向荷载，而两边格构柱承担较大的弯矩，施工过程中的弯矩大约为 500kN·m。研究成果为“一柱三桩”的优化设计和数值模拟研究提供支撑。

关键词：逆作法；一柱三桩；结构托换；自动化监测

Research on vertical load transfer law of " One Column Three Piles" system with top-down construction method

Deng Liang[1,2]

(1. Shanghai Construction No. 2 (Group) Co., Ltd., Shanghai 200080, China;
2. Shanghai Engineering Research Center of Top-Down Method in Construction Engineering, Shanghai 200080, China)

Abstract: In order to study the mechanics response law of the " One Column Three Piles" system, the self-developed automatic monitoring system was used during the construction of the 4# building of the Hongqiao Imported Commodity Exhibition and Trading Center (Phase Ⅱ). The vertical bearing structure of the “One Column Three Piles” system is monitored in real time. The monitoring results show: 1) During the construction process of the upper and lower synchronous reverse construction method, the vertical structure of the " One Column Three Piles" system on B0 board has periodic " tension and compression conversion", which has a causal relationship with the excavation conditions; 2) During the construction of the " One Column Three Piles" system, the steel column of B0 board bears about 50% of the vertical load, while the lattice columns on both sides bear a larger bending moment. The bending moment during the construction process is about 500kN·

m. The research results provide support for the optimization design and numerical simulation research of "One Column Three Piles" system.
Keywords: top-down construction method; "One Column Three Piles" system; structure underpinning; automatic monitoring system

装配式建筑无外脚手架施工的防高坠管理研究

马跃强　李卫红　陆冬兴　刘成涛　李　锋
（上海建工二建集团有限公司，上海，200080）

摘　要：以上海市三林镇0901-13-02号地块项目为研究对象，探索装配式建筑在无外脚手架施工方法下的防高坠安全管理措施。经过自主研发各类防坠落安全设施，采用模拟实验并实际应用，确保了工程快速作业下的施工安全。本文对项目中既有特殊性、又有普遍性的防高坠安全防护措施的具体做法进行介绍，对施工经验进行总结。
关键词：装配式建筑；无外脚手架施工；水平外挑网

Research on anti falling management of prefabricated building without external scaffold construction

MaYueqiang　Li Weihong　Lu Dongxing　Liu Chengtao　Li Feng
（Shanghai Construction No. 2（Group）Co. , Ltd. , Shanghai 200080, China）

Abstract: Taking the project of plot 0901-13-02 in Sanlin Town of Shanghai as the research object, this paper explores the safety management measures for preventing high falling of prefabricated buildings under the construction method without external scaffolding. Through independent research and development of all kinds of fall prevention safety facilities, using simulation experiments and practical application, the construction safety under the rapid operation of the project is ensured. This paper introduces the specific methods of the safety protection measures against high falling, which have both particularity and universality in the project, and summarizes the construction experience.
Keywords: prefabricated building; no external scaffold construction; horizontal external net

路面加筋网及封层在反射裂缝防治中的应用

徐尤旺
（中国电建集团江西省水电工程局有限公司，江西 南昌，330000）

摘　要：依托工程实践，通过前期的病害分析，在对比分析不同加筋材料防治反射裂缝效果的基础上，选定了道路加铺结构设计方案，采用钢丝网+1cm稀浆封层的综合治理防治减缓“白改黑”反射裂缝。经改造后的沥青道路运营7个月后，沥青路面未出现反射裂缝、推移、车辙等病害，行车舒适、平稳，有效提高了道路使用性能及耐久性，表明加筋网施工方便灵活、施工速度快，沥青封层工艺成熟、可操作性强，同时具有良好的经济效果，值得推广。

关键词：加筋网；封层；反射裂缝；防治；白改黑

Application of pavement reinforced Mesh and sealing layer in prevention and cure of reflection crack

Xu Youwang
（Powerchina Jiangxi Hydropower Engineering Bureau Co.，Ltd.，
Nanchang，Jiangxi 330000，China）

Abstract：Relying on engineering practicet，through preliminary investigation about the road and analyze the effect about different reinforcement materials in crack control. The design scheme of road overlay structure is selected，Galvanized Mesh and Slurry seal of 1cm. Open traffic for seven months in asphalt road，there is no reflection crack，displacement，rutting and other diseases on asphalt pavement，driving comfortable，smooth，effectively improve the road performance and durability，resulting in better economic and social benefits. The results show that the construction of zinc-plated reinforcing net is convenient and flexible，the speed is fast，the technology of asphalt sealing coat is mature，the Operability is easy，and the economic effect is good，which is worth popularizing.

Keywords：reinforced mesh；seal；reflection crack；prevention and cure；white to black

工程结构超载设计的极限状态分析

黄卓驹[1,2]
（1. 同济大学建筑设计研究院（集团）有限公司，上海，200092；
2. 同济大学 建筑工程系，上海，200092 ）

摘　要：超载现象在工程中不罕见，由于现实中的种种原因，结构最终使用的外部荷载或作用可能超出预定水平，当前这种风险很难预估。实践中很多设计者会简单采用提高标准荷载水平的方法进行超载设计，本文系统地对这种设计带来的影响进行讨论。首先建立了工程结构设计的数学模型，通过几何分析的工具对极限状态进行了研究，对不同的结构-荷载对效应/抗力的影响情况分类讨论，进行了直观的定性分析。并对非确定情况进行了初步的讨论。由此得出提高荷载标准进行设计是不可靠的，应从效应和抗力关系入手增加结构的冗余度，或考虑独立的超载工况组合进行包络设计的方式，才能真正有效提高设计可靠性的结论。

关键词：结构设计；可靠性；超载；极限状态

Limit State Analysis on Overload Design for Engineering Structure

Huang Zhuoju[1,2]
(1. Tongji Architectural Design (Group) Co., Ltd., Shanghai 200092, China;
2. Department of Structural Engineering, Tongji University, Shanghai 200092, China)

Abstract: Overloading is common in structural engineering, and for a variety of reasons in reality, the external loads or actions that a structure may end up using may exceed the intended levels, a risk that is currently difficult to predict. In practice many designers will simply adopt the method of increasing the standard value of the load for overload design, and the implications of such design are systematically discussed in this paper. Firstly, a mathematical model for the design of engineering structures is developed, the limit states are investigated by means of the tools of geometric analysis, and the effects of different structural-loads on the effect/resistance cases are categorized and discussed in a visual qualitative analysis. A preliminary discussion of the non-deterministic cases is also presented. This leads to the conclusion that raising the load criteria for design is unreliable and that the redundancy of the structure should be increased by starting with the effect and resistance relationship or by considering independent combinations of overload conditions for envelope design in order to truly and effectively improve design reliability.

Keywords: structural design; reliability; overload; limit state